# ARITHMETIC

## AN INTRODUCTION TO MATHEMATICS

*Bevan K Youse*
*Emory University*

*CANFIELD PRESS • SAN FRANCISCO*
*A Department of Harper & Row, Publishers, Inc.*
*New York         Evanston         London*

ARITHMETIC: AN INTRODUCTION TO MATHEMATICS
Copyright © 1971 by Bevan K Youse

Printed in the United States of America. All rights reserved. No part of this book may be used or reproduced in any manner whatsoever without written permission except in the case of brief quotations embodied in critical articles and reviews. For information address Harper & Row, Publishers, Inc., 49 East 33rd Street, New York, N.Y. 10016.

Standard Book Number: 06–389650–8

Library of Congress Catalog Card Number: 76–143695

# CONTENTS

Preface  *vii*

Mathematical Symbols and Their Meaning  *x*

## 1. Numbers and Numerals   3
    1.1  The Number Concept  *3*
    1.2  Notational Systems for Numbers  *6*
    1.3  Sets and Set Operations  *11*
    1.4  Supplementary Exercises  *17*

## 2. Addition of Whole Numbers   21
    2.1  Addition  *21*
    2.2  Techniques for Addition  *28*
    2.3  Checks for Addition  *33*
    2.4  Supplementary Exercises  *35*

## 3. Subtraction of Whole Numbers   37
    3.1  Subtraction  *37*
    3.2  Techniques for Subtraction  *40*
    3.3  Less Than and Greater Than  *43*
    3.4  Supplementary Exercises  *45*

## 4. Multiplication of Whole Numbers   49
    4.1  Multiplication  *49*
    4.2  Techniques for Multiplication  *55*
    4.3  Factors  *59*
    4.4  Supplementary Exercises  *62*

## 5. Division and Incomplete Quotients  65
   5.1  Division and Divisibility Tests  *65*
   5.2  The Division Algorithm  *71*
   5.3  Other Multiplication Algorithms  *76*
   5.4  Supplementary Exercises  *80*

## 6. Fundamental Theorem of Arithmetic  83
   6.1  Primes  *83*
   6.2  Fundamental Theorem of Arithmetic  *87*
   6.3  G.C.F. and L.C.M.  *89*
   6.4  Euclidean Algorithm  *94*
   6.5  Supplementary Exercises  *98*

## 7. The Rational Numbers  101
   7.1  The Rational Numbers  *101*
   7.2  Equality and Inequality  *108*
   7.3  Supplementary Exercises  *112*

## 8. Addition and Subtraction of Rational Numbers  115
   8.1  Fractions  *115*
   8.2  Mixed Numerals  *120*
   8.3  Decimals  *124*
   8.4  Approximating Rationals with Decimals  *129*
   8.5  Supplementary Exercises  *135*

## 9. Multiplication and Division of Rational Numbers  137
   9.1  Fractions  *137*
   9.2  Decimals  *143*
   9.3  Complex Fractions and Complex Decimals  *147*
   9.4  Supplementary Exercises  *151*

## 10. The Negative Integers and Rationals  153
   10.1  Extending the Number Line  *153*
   10.2  Multiplication and Division  *158*
   10.3  Inequalities  *161*
   10.4  Supplementary Exercises  *162*

Contents

## 11. Applications  165
    11.1 Introduction  *165*
    11.2 Linear Measure  *166*
    11.3 Area  *170*
    11.4 Volume  *176*
    11.5 Other Measures  *180*
    11.6 Miscellaneous Applications  *185*
    11.7 Arithmetic Sequences  *189*
    11.8 Geometric Sequences  *196*

## 12. Basic Theorems  203
    12.1 Divisibility Tests  *203*
    12.2 The Infinitude of Primes  *207*
    12.3 Fundamental Theorem of Arithmetic  *210*
    12.4 Euclidean Algorithm  *214*
    12.5 Number Congruence  *220*
    12.6 Other Number Bases  *227*
    12.7 Infinite Decimals  *235*
    12.8 Irrational Numbers  *240*

    Mathematical Terms and Basic Properties  *245*
    Answers to Odd-Numbered Exercise Problems  *251*
    Compound Interest Tables  *283*
    Index  *291*

# PREFACE

This book is designed for the student seeking some combination of the following: (1) a review of the basic concepts and techniques of arithmetic, (2) a discussion of the applications of arithmetic and the techniques for solving practical problems, (3) an understanding of why the algorithms of arithmetic produce the correct results, and (4) a knowledge of the underlying principles and basic theory of arithmetic.

Every effort is made to present the material in such a way that a proper balance can be maintained between an emphasis on computational skills and on the theory and structure of the number system. One feature of the text which helps attain the desired balance is the inclusion of two exercise sets at the end of nearly every section. The second of these exercise sets generally includes problems which require more sophisticated computational skills or which lead to a better understanding of the theory of arithmetic. The two types of exercise sets at the end of each section, the separate chapter on applications of arithmetic, and the final chapter presenting the basic theory of arithmetic should make it possible for the instructor to tailor a course to fit both the needs and the personality of the class.

It is expected that many of the carefully chosen problems will promote some enthusiasm for doing much of the drill work necessary to learn computational techniques. The student will discover that many interesting problems can be solved by computation alone while some will require making conjectures and seeking answers to thought-provoking questions. At the end of most chapters is a section with supplementary exercises: these should give the student any necessary additional practice in computation as well as an adequate review of the ideas and techniques discussed in the chapter.

Only those terms which seem essential and important to an understanding of arithmetic have been introduced. The fundamental terms are prominently displayed and should be learned by the student since they are of vital importance in the communication of the basic ideas. To aid the student, each new term is generally used several times in the text soon after it is introduced.

A serious look has been taken at what seems to be the essential techniques and the basic theory of arithmetic. As a result, a few topics of tangential interest usually discussed in a traditional approach to arithmetic have been omitted as well as some of the more sophisticated topics from the most modern approach to the subject.

Chapter 1 discusses numbers and numerals and the basic set concepts which are essential to a clear understanding of arithmetic. Only that theory of sets which is of importance to arithmetic is discussed. A careful distinction is made between numbers, names for numbers, and the notational systems used to denote numbers.

In Chapter 2, *Addition of Whole Numbers*, the set of whole numbers is introduced. The operation of addition on the set of whole numbers is defined and several techniques for addition are discussed as well as methods of check for addition.

In Chapter 3, subtraction is defined and techniques for performing this operation are presented. Also, the greater-than and less-than relations are defined and the fundamental properties of inequality are discussed.

In Chapter 4, *Multiplication of Whole Numbers*, the operation of multiplication on the set of whole numbers is defined and several techniques for finding products are presented. The important ideas of factor and multiple are also introduced in this chapter.

In Chapter 5, the fourth rational operation, division, is defined. At this point, several divisibility tests are given and discussed. The division algorithm is also presented and a careful distinction is made between the rational operation of division and the long division algorithm used for finding incomplete quotients and remainders.

Chapter 6, *Fundamental Theorem of Arithmetic*, is devoted to an elementary discussion of prime numbers. Also discussed are the Fundamental Theorem of Arithmetic, the Euclidean algorithm, greatest common factor, and least common multiple. Although proofs of the Fundamental Theorem and the Euclidean algorithm are postponed until Chapter 12, their importance to the development of arithmetic is discussed at this time.

*Preface*

The rational numbers are defined in Chapter 7 and the fractional notation for these numbers is presented. The properties of equality and inequality are also explained.

In Chapter 8, the operations of addition and subtraction for rational numbers are studied. The mixed numeral notation and the decimal notation for rational numbers are introduced as well as an algorithm for finding approximations in decimal form for any rational number.

Definitions of multiplication and division on the set of rational numbers are presented in Chapter 9. Complex fractions and complex decimals are also introduced.

Chapter 10 gives a discussion of the negative integers and negative rational numbers. The techniques for performing the rational operations on the set of all rational numbers is given and the concept of inequality is extended to the rational number system.

Chapter 11 presents an extensive discussion of applications of arithmetic. Some of the topics discussed are linear measure, area, volume, percent, and interest.

In Chapter 12, the divisibility tests are restated and the validity of each is proved. It is also proved that the set of prime numbers is an infinite set. The Fundamental Theorem of Arithmetic and the Euclidean algorithm are restated and proved. Number congruence is defined and the standard theorems associated with this important concept are stated and proved. Number bases other than ten are developed including the binary system, base six, and base eight. Finally, a short discussion is included on infinite decimals and irrational numbers. This chapter is of particular interest to the student who needs a clear understanding of the theoretical aspects of arithmetic.

I am particularly indebted to Professor Dora Helen Skypek for reading the manuscript and making many valuable suggestions for improvement. The author is also indebted to Prentice-Hall, Inc., for permission to include material taken from his book *Arithmetic: A Modern Approach*, 1963, of which they are the publishers.

Bevan K Youse

# MATHEMATICAL SYMBOLS AND THEIR MEANINGS

| | |
|---|---|
| $\in$ | Is an element in. |
| $\notin$ | Is not an element in. |
| $a \in S$ | Element $a$ is in set $S$. |
| $a \notin S$ | Element $a$ is not in set $S$. |
| $S \cap T$ | Intersection of sets $S$ and $T$. |
| $S \cup T$ | Union of sets $S$ and $T$. |
| $S'_T$ | Complement of set $S$ relative to set $T$. |
| $\emptyset$ | Empty set. |
| $n(S)$ | Number of elements in set $S$. |
| $+$ | Addition. |
| $-$ | Subtraction. |
| $\times$ | Multiplication. |
| $\div$ | Division. |
| $=$ | Equality. |
| $<$ | Less-than. |
| $>$ | Greater-than. |
| $\leq$ | Less-than or equal to. |
| $\geq$ | Greater-than or equal to. |
| $a\|b$ | $a$ is a factor of $b$. |
| $d(n)$ | Number of positive factors of a natural number $n$. |
| $s(n)$ | Sum of all the positive factors of a natural number $n$. |
| $\phi(n)$ | Number of positive integers less than $n$ which are relatively prime to $n$. |

**1**

**1234567890**
IIIIIIIVVVIVIIVIIIIXXXIXIIXIIIXIVXV

**IV**
12345
12345678
IIIIIIIIIIIII

**X**

MCMLXXI1971

# NUMBERS AND NUMERALS

## 1.1 The Number Concept

The development of a number system from the rudimentary number concept possessed by early man ranks with the development of language as one of man's greatest and most distinctive creations. In the study of arithmetic, we learn the basic features of this important invention. We discuss (a) the notations for numbers, (b) the elementary operations performed on numbers, (c) the techniques for performing these operations, and (d) the use and applications of numbers in everyday pursuits.

Basic to the number concept is man's ability to perceive the difference in sizes of sets; that is, an ability to distinguish the set ∗ ∗ ∗ ∗ ∗ ∗ from the set ∗ ∗ ∗ ∗. We use the word "six" to describe the size of the first set and the word "four" to describe the size of the second set. The first set is said to have six as its cardinal number. Cardinal number answers the question, how many?, and indicates the size of a set.

### Examples

1. The cardinal number of the set of states in the United States of America is fifty.
2. Eighty-eight is the cardinal number of the set of keys on a standard piano.

We know that the words (sounds) for numbers vary with the languages developed by man. For example, today the Germans use *sechs*, the Spanish use *seis*, and we use *six* to name the same number.

However, in these countries, the symbols and basic features of the number system are practically the same. In this sense, arithmetic is truly an international language. As we shall observe shortly, although the number concept does not depend on language, the structure of the number system is reflected in the words for numbers.

It was not a trivial accomplishment when man first observed the common property possessed by sets such as pictured in Figure 1. Although the sets contain different elements, the elements in any one set can be paired in a one-to-one fashion with the elements in each of the other sets. In the early development of the idea of numbers, man used his fingers to pair with elements in the sets with which he dealt. Even today most persons have used their fingers at one time or another for doing arithmetic computations.

The fact that we have ten fingers is the reason why our present number system uses ten as its base number. The base number is the number of objects we put into basic groupings for the purposes of counting. By considering the words we use for the first thirty counting numbers, it will be obvious how the names chosen for numbers have been influenced by the base number.

| one   | eleven (one and ten)      | twenty-one (two tens and one)     |
|-------|---------------------------|-----------------------------------|
| two   | twelve (two and ten)      | twenty-two (two tens and two)     |
| three | thirteen (three and ten)  | twenty-three (two tens and three) |
| four  | fourteen (four and ten)   | twenty-four (two tens and four)   |
| five  | fifteen (five and ten)    | twenty-five (two tens and five)   |
| six   | sixteen (six and ten)     | twenty-six (two tens and six)     |
| seven | seventeen (seven and ten) | twenty-seven (two tens and seven) |
| eight | eighteen (eight and ten)  | twenty-eight (two tens and eight) |
| nine  | nineteen (nine and ten)   | twenty-nine (two tens and nine)   |
| ten   | twenty (two tens)         | thirty (three tens)               |

Nearly all words for the counting numbers clearly indicate the use of ten for grouping. The only exceptions in the above list are the words "eleven" and "twelve"; the natural choices for these numbers would seem to be "oneteen" and "twoteen." The historical reason for having the twelve distinct names is probably the fact that man has also used twelve as a base number. Although twelve is not the base number for our present-day number system, grouping by twelves still permeates our daily activities. For example, there are twelve eggs in a dozen, twelve groups of twelve objects in a gross, and twelve inches in a foot.

## 1.1 The Number Concept

Figure 1.  *The number concept two.*

When we say that a student is sixth in his class, the number is being used in an *ordinal sense*; that is, the students have been ordered in some prescribed way, such as by age, height, grade average, weight, or I.Q., and we assert that the student occupies position six in the listing. Ordinal numbers signify the positions occupied by the elements of a set after they have been lined up in some linear arrangement.

### Examples

In each of the following, the numbers are used in an ordinal sense.
1. John sits in the sixth seat in row three.
2. He is number one in his class.

Today, the counting numbers—one, two, three, four, five, etc.—are often called the natural numbers. The natural numbers served man's needs quite well for a great many years and the extension of the number system to include the familiar number "zero" took several centuries. From a modern point of view, zero is the answer to the question, how many?, after all objects from a given set have been removed.

The numbers—zero, one, two, three, four, etc.—consisting of the natural numbers and zero are usually called the whole numbers. In the next section we shall discuss the symbols and the system of notation used for this important set of numbers.

Although it may not be obvious, the introduction of zero into the number system required a much higher level of sophistication than was needed for the development of the natural numbers. The level of abstraction necessary and the inherent difficulties will be evidenced by the special treatment required for zero as we discuss the development of our number system. The student should make certain to observe the distinction between the set of natural numbers and the set of whole numbers. Because of the special properties of zero, a statement which is true for one set is often not true for the other set.

### Exercises I

1. List four things which we study in arithmetic.
2. What is a cardinal number?
3. What is the base number of our number system?
4. What does the base number indicate in terms of counting?
5. What is an ordinal number?
6. What is another name for the set of counting numbers?
7. What numbers are in the set of whole numbers?
8. What names do we use for each of the following groupings?
   (a) Five tens.   (b) Eight tens.   (c) Nine tens.
   (d) Ten tens.   (e) Ten groups of ten tens.

In Exercises 9 through 12, indicate in each case whether the number is being used in the ordinal or cardinal sense.

9. The elements in a set with six elements cannot be paired in a one-to-one fashion with the elements in a set with four elements.
10. In a class of twenty-five students, John ranks seventh.
11. There are 18 pages in this chapter and this problem set is is on page 6.
12. There are three persons living at two-one-five Johnson Avenue.

### Exercises II

1. List all the English words you can think of that express the idea of *twoness*.

*1.2 Notational Systems for Numbers* 7

2. List all the English words you can think of that express the idea of *threeness*.
3. The ancient Babylonians used sixty as a base number. What evidences of this base number can you find in present-day groupings?
4. Explain how we can distinguish between the set * * * * and the set * * * * * * without resorting to counting or using numbers.
5. Suppose that *un, do, to, fo*, and *dot* were the words used for the first five counting numbers and suppose we used five as the base number. If *un, do, to, fo, dot, undot, dodot, todot, fodot, dodoty, dodoty-un* were the names of the first eleven numbers, what do you think might be the names for the next thirteen counting numbers in this system?

If *dooty* were the name given to five groups of five in the system with base five as described in the previous exercise, state the standard name in English for each of the numbers in Exercises 6 through 12.

6. (a) Un dooty and un.
   (b) Un dooty and dot.
7. (a) Un dooty and dodoty.
   (b) Un dooty and fo.
8. (a) Un dooty and fodoty.
   (b) Un dooty and dodoty-do.
9. (a) Do dooty and dodoty.
   (b) Do dooty and dodoty-do.
10. (a) Do dooty and todoty.
    (b) Fo dooty and fodoty-fo.
11. (a) To dooty and to.
    (b) Fo dooty and todot.
12. (a) To dooty and fodot.
    (b) To dooty and todoty-do.

## 1.2 Notational Systems for Numbers

The basic symbols used as names for numbers are called numerals; they are also often called digits. The numerals we use are

$$0, 1, 2, 3, 4, 5, 6, 7, 8, 9.$$

These numerals, as well as our notational system, were first developed by the Hindus. Later, the Arabs adopted the system from the Hindus, contributed to its further development, and introduced it into Europe. Because of its origin, we call our present system the Hindu–Arabic number system. We should observe that the number of numerals in the Hindu–Arabic system is the same as the base number.

The Romans used a different set of numerals to denote numbers. The Roman Numerals are

I, V, X, L, C, D, M

where "I" denotes one, "V" denotes five, "X" denotes ten, "L" denotes fifty, "C" denotes one hundred, "D" denotes five hundred, and "M" denotes one thousand. As late as the seventeenth century, these numerals were still in prominent use in Europe for the purpose of keeping records.

Not only are the numerals in the Roman system different from the numerals in the Hindu–Arabic system, but also the two notational systems are quite different in another way. Since positions in the Roman notation are not reserved for basic groupings as in the Hindu–Arabic system, the Roman system is not a positional system in the sense that the Hindu–Arabic system is. With Roman numerals, "VI" denotes six (five plus one) and "IV" denotes four (five less one). This second difference in the two systems is made quite obvious by just considering "51" and "15" in the Hindu–Arabic notation.

The way in which Roman numerals are used to denote numbers makes it unnecessary to have a numeral for zero. On page 9 are several examples of Roman numeral designations for natural numbers. These numerals exhibit the basic principles underlying the Roman notational system for natural numbers.

In the Hindu–Arabic notation, one thousand seven hundred thirty-six is denoted by 1736, or 1,736. The numeral "1" occupies what is called the *thousands'* (ten hundreds) position, "7" occupies the *hundreds position*, "3" is in the *tens' position*, and "6" is in the *units' position*, or *ones' position*. This positional notation with ten as a base number uses "0" as a *place-holder*. The numeral 0 is used to occupy certain positions and act as a spacer so that we are able to distinguish between such numbers as 34 (thirty-four) and 3,004 (three thousand four). In Figure 2, we indicate the names of the first several positions. We see that after the first three positions we invent new names only for each successive set of three positions. Each of the groups of three positions is called a period, and each period is generally set off by commas. The names of the first five

## 1.2 Notational Systems for Numbers

| I | one | XI | eleven | XXX | thirty |
|---|---|---|---|---|---|
| II | two | XII | twelve | XXXV | thirty-five |
| III | three | XIII | thirteen | XL | forty |
| IV | four | XIV | fourteen | XLV | forty-five |
| V | five | XV | fifteen | LI | fifty-one |
| VI | six | XVI | sixteen | LX | sixty |
| VII | seven | XVII | seventeen | LXIII | sixty-three |
| VIII | eight | XVIII | eighteen | LXX | seventy |
| IX | nine | XIX | nineteen | LXXXI | eighty-one |
| X | ten | XX | twenty | XCV | ninety-five |

| | |
|---|---|
| CCIV | two hundred four |
| CCLXXIV | two hundred seventy-four |
| CCXCII | two hundred ninety-two |
| CMXCIII | nine hundred ninety-three |
| MMC | two thousand one hundred |
| MCDXCII | one thousand four hundred ninety-two |
| MDCCLXXVI | one thousand seven hundred seventy-six |

periods starting at the right and reading to the left are *ones* (or *units*), *thousands*, *millions*, *billions*, and *trillions*.

The base number for a positional notation is quite arbitrary. If man had developed eight fingers instead of ten, we would probably be using eight as our base number. It might have been better if we had developed twelve fingers since a number system with twelve as a base number has some slight advantages over base-ten. In

| PLACE-VALUES | Hundred trillions | Ten trillions | Trillions | Hundred billions | Ten billions | Billions | Hundred millions | Ten millions | Millions | Hundred thousands | Ten thousands | Thousands | Hundreds | Tens | Ones |
|---|---|---|---|---|---|---|---|---|---|---|---|---|---|---|---|
| NUMBER | 2 | 6 | 1, | 4 | 2 | 7, | 6 | 2 | 1, | 4 | 6 | 3, | 8 | 2 | 6 |
| PERIODS | Trillions ||| Billions ||| Millions ||| Thousands ||| Ones |||
| NAME | Two hundred sixty-one trillion, four hundred twenty-seven billion, six hundred twenty-one million, four hundred sixty-three thousand, eight hundred twenty-six. ||||||||||||||| |

Figure 2

fact, some people have recommended making the change to the *duodecimal* (base-twelve) system. However, since the advantages are so slight and our base-ten system is so entrenched, there seems little chance that this change would be considered seriously by enough persons to effect the change.

In Chapter 12, we shall discuss in detail different number bases and indicate various advantages and disadvantages in using other base numbers. As we develop the arithmetic for the Hindu–Arabic system, it would be worthwhile for the student to observe that there is nothing "sacred" about ten as a base number and that two, eight, or twelve, for example, could be used just as well.

### Exercises I

1. (a) List the Hindu–Arabic numerals.
   (b) List the Roman numerals.
2. State in words the name of each of the following numbers.
   (a) 2,684. (b) 48,206. (c) 60,002.
3. State in words the name of each of the following numbers.
   (a) 126,820. (b) 3,106,406. (c) 12,000,041.
4. State in words the name of each of the following numbers.
   (a) XXII. (b) XLVI. (c) CCXXIX.
5. State in words the name of each of the following numbers.
   (a) XLIV. (b) MCCLXV. (c) MCMLXXII.

In Exercises 6 through 10, write each number in both Hindu–Arabic and Roman notation.

6. (a) Fifty-eight. (b) One hundred six.
7. (a) Forty-eight. (b) One hundred thirty-one.
8. (a) Eighty-nine. (b) One thousand sixty-six.
9. (a) Ninety-four. (b) Fourteen hundred ninety-two.
10. (a) One thousand three. (b) Nineteen hundred seventy.

### Exercises II

1. As stated earlier, the numerals 0, 1, 2, 3, 4, 5, 6, 7, 8, 9 are also called *digits*. Look up the origin of the word "digit" in a dictionary.

## 1.3 Sets and Set Operations

2. We use numbers for purposes of calculation. Look up the origin of the word "calculate" in a dictionary.
3. For the number system with five as base number which we discussed in the second set of exercises in the last section, let ⇂, ℨ, Ɛ, ƕ be the numerals for *un, do, to, fo,* respectively. If a positional notation is constructed with these numerals and with five as base number, the first eleven numbers would be denoted as follows.

⇂, ℨ, Ɛ, ƕ, 10, 11, 1ℨ, 1Ɛ, 1ƕ, ℨ0, ℨ⇂.

Use this notation to write the next fifteen numbers.

Use the numerals given in Exercise 3 to name each indicated number from the second set of exercises in Section 1.1.

4. Exercise 6.  5. Exercise 7.  6. Exercise 8.
7. Exercise 9.  8. Exercise 10.  9. Exercise 11.
10. Write each of the following numbers in Roman notation and in Hindu–Arabic notation. (a) Three hundred four. (b) Nine hundred forty-six.

State in words the name of each number in the following statements.

11. There are 3,600 seconds in an hour, 86,400 seconds in a day, and 41,536,000 seconds in a (non-leap) year.
12. There are 373,824,000,000 seconds in 9,000 (non-leap) years.
13. The national debt of the United States in early 1971 was in excess of 373,000,000,000 dollars.
14. The closest distance from earth to the moon is approximately 221,463 miles.
15. The distance from the earth to the sun is approximately 93,000,000 miles.
16. The distance to Mars from the sun is approximately 141,710,000 miles.
17. The distance to Pluto from the sun is approximately 3,675,270,000 miles.

## 1.3 Sets and Set Operations

Sets and related ideas are quite valuable in the development of arithmetic. Therefore, let us turn our attention to some of the basic set concepts before beginning a discussion of the arithmetic operations and the techniques for performing these operations.

When we consider, for example, the set consisting of the elements $a$, $b$, and $c$, it is usually denoted by $\{a, b, c\}$. In general, a pair of braces $\{\ \}$ is used to enclose the elements of a set which we wish to define. Since a set is determined by its members, the order of listing the members is immaterial; thus, $\{a, b, c\}$ and $\{b, c, a\}$ are the same sets. We say that the two sets are *equal* and write $\{a, b, c\} = \{b, c, a\}$ to state this fact.

We often use letters (usually capitals) for the names of sets. For example, we might let $S$ be the set containing the elements $a$, $b$, and $c$; that is, we let $S = \{a, b, c\}$. To express the fact that $a$ is an element in $S$ we write $a \in S$ (read "$a$ is in $S$"). Similarly, $b \in S$ and $c \in S$. Since $d$, for example, is not in $S$, we write $d \notin S$ (read "$d$ is not in $S$") to state this fact.

### *Example*

Let $S = \{a, b, c\}$ and $T = \{b, c, d, e\}$. Then, $a \in S$, $a \notin T$, $b \in S$, $b \in T$, $c \in S$, $c \in T$, $d \notin S$, $d \in T$, $e \notin S$, $e \in T$.

If $S = \{a, b, c\}$ and $T = \{b, c, d, e\}$, then the set $\{a, b, c, d, e\}$ which consists of the elements in $S$, in $T$, or in both $S$ and $T$ is called the union of the given sets. The set is denoted by $S \cup T$ (read "$S$ union $T$"). Thus,

$$S \cup T = \{a, b, c, d, e\}.$$

Since the union of two sets is a set, union is called a binary operation. For sets, a binary operation is a prescribed method by which we construct a new set from *two* given sets.

### *Examples*

1. If $T = \{b, c, d, e\}$ and $W = \{p, q, r\}$, then
$$T \cup W = \{b, c, d, e, p, q, r\}.$$
2. If $T = \{b, c, d, e\}$ and $H = \{c, d\}$, then
$$T \cup H = \{b, c, d, e\} = T.$$

If $S = \{a, b, c\}$ and $T = \{b, c, d, e\}$ then the set $\{b, c\}$ which consists of the elements in *both* $S$ and $T$ is called the intersection of the given sets. This is denoted by $S \cap T$ (read "$S$ intersection $T$"). Thus,

$$S \cap T = \{b, c\}.$$

## 1.3 Sets and Set Operations

### Examples

1. If $V = \{q, r, s\}$ and $W = \{p, q, r\}$, then
$$V \cap W = \{q, r\}.$$

2. If $T = \{b, c, d, e\}$ and $H = \{c, d\}$, then
$$T \cap H = \{c, d\} = H.$$

We have already indicated that we admit the existence of an *empty set*, a set with no elements. One advantage of having an empty set is that it ensures the intersection of any two sets is a set; thus, intersection is also a binary operation defined on sets. If $T = \{b, c, d, e\}$ and $S = \{p, q, r\}$, then $S \cap T$ is the empty set since these two sets have no elements in common. We use the symbol "$\emptyset$" to denote the empty set. Two sets, such as $S$ and $T$, which have no elements in common are said to be disjoint; that is, if $S \cap T = \emptyset$, then $S$ and $T$ are disjoint sets.

In mathematics, parentheses are often used to indicate the order in which operations are to be performed. If $A$, $B$, and $C$ are sets, then $A \cap (B \cup C)$ is the intersection of $A$ with the union of $B$ and $C$. Similarly, $(A \cap B) \cup C$ is the union of the intersection of $A$ and $B$ with the set $C$. One example will show that, in general, these two sets are not equal.

### Example

Let $A = \{a, b, c, d\}$, let $B = \{c, d, e, f\}$, and let $C = \{a, d, h, j\}$. Then,

$$B \cup C = \{a, c, d, e, f, h, j\} \quad \text{and} \quad A \cap (B \cup C) = \{a, d, c\}.$$

Also,

$$A \cap B = \{c, d\} \quad \text{and} \quad (A \cap B) \cup C = \{a, c, d, h, j\}.$$

Let $A$ and $B$ be any sets. It should be obvious that $A \cup B$ and $B \cup A$ are the same. We say that union is a commutative operation to indicate that

$$A \cup B = B \cup A$$

for any two given sets $A$ and $B$. Also, a few examples should convince us that $(A \cup B) \cup C$ and $A \cup (B \cup C)$ are the same sets for any sets $A$, $B$, and $C$. We say that union is an associative operation to indicate that

$$A \cup (B \cup C) = (A \cup B) \cup C.$$

Since the order of performing the union operation on sets is immaterial, we often omit the parentheses and just write $A \cup B \cup C$ for the union of the given sets. A few examples should also convince us that intersection is both a commutative and an associative operation; that is, for sets $A$, $B$, and $C$,

$$A \cap B = B \cap A \quad \text{and} \quad A \cap (B \cap C) = (A \cap B) \cap C.$$

## Examples

Let $A = \{a, b, c, d, e\}$, let $B = \{c, d, e, f, g\}$, and let $C = \{d, e, f, g, h\}$.
1. $A \cup B = \{a, b, c, d, e, f, g\}$ and $B \cup C = \{c, d, e, f, g, h\}$. Thus,

$$(A \cup B) \cup C = \{a, b, c, d, e, f, g, h\}$$

and

$$A \cup (B \cup C) = \{a, b, c, d, e, f, g, h\}.$$

2. $A \cap B = \{c, d, e\}$ and $B \cap C = \{d, e, f, g\}$. Thus,

$$(A \cap B) \cap C = \{d, e\}$$

and

$$A \cap (B \cap C) = \{d, e\}.$$

If $S = \{a, b, c\}$ and $R = \{a, b, c, d, e\}$, then not only do the sets have elements in common but every element in $S$ is in $R$. In this case, we say that $S$ is a subset of $R$ and denote this fact by $S \subseteq R$ (read "$S$ is a subset of $R$"). In general, a set $S$ is called a subset of $R$ if and only if every element in $S$ is in $R$. As another example, $\{1, 2\} \subseteq \{1, 2, 3, 4\}$.

For any set $S$, we know that $S \subseteq S$ since every element in $S$ is in $S$. Not only is any set $S$ a subset of itself but also the empty set is a subset of any given set; that is, $\emptyset \subseteq S$.

As stated above, if $S = \{a, b, c\}$ and $R = \{a, b, c, d, e\}$, then $S \subseteq R$. Often for two sets such as these, we are interested in the set of elements in $R$ which are *not in* $S$; in our example, the set is $\{d, e\}$. In general, if $S \subseteq R$ then the set consisting of all elements in $R$ which are not in $S$ is called the complement of $S$ relative to $R$; we denote it by $S'_R$ (read "complement of $S$ relative to $R$"). If there

## 1.3 Sets and Set Operations

is no question as to what set the complement is taken relative to, we write $S'$ instead of $S'_R$ for the set and call it the complement of $S$.

### Examples

Let $S$ be the set of students in a given class, let $B$ be the set of boys in the class, and let $G$ be the set of girls in the class.
1. $B \subseteq S$ and $G \subseteq S$.
2. $B \cap G = \emptyset$, the sets $B$ and $G$ are disjoint.
3. $B \cup G = S$.
4. $B'_S$, or $B'$, is the set of students in the class which are not boys; thus, $B' = G$. Similarly, $G' = B$.

Let $S = \{a, b, c\}$. We use the symbol $n(S)$ (read "$n$ of $S$") to denote the number of elements in $S$. For the given set, $n(S) = 3$. If $H$ is the set of Hindu–Arabic numerals, then $n(H) = 10$. Since the cardinal number of the empty set is zero, $n(\emptyset) = 0$.

We have already seen that the pairing of objects in disjoint sets is basic to man's number concept. Two sets for which the elements can be paired in a one-to-one fashion are called equivalent sets. Equivalent sets are the *same size* and have the *same cardinal number*. A one-to-one pairing is often called a one-to-one correspondence. The way in which two equivalent sets, such as $\{a, b, c\}$ and $\{e, f, g\}$, can be put into one-to-one correspondence is not unique; that is, the pairing can be done in several ways. For the two given sets with three elements each, there are six different one-to-one correspondences. They are exhibited as follows.

ONE-TO-ONE CORRESPONDENCES BETWEEN $\{a, b, c\}$ AND $\{e, f, g\}$

| $a \leftrightarrow e$ | $a \leftrightarrow e$ | $a \leftrightarrow f$ |
|---|---|---|
| $b \leftrightarrow f$ | $b \leftrightarrow g$ | $b \leftrightarrow e$ |
| $c \leftrightarrow g$ | $c \leftrightarrow f$ | $c \leftrightarrow g$ |
| $a \leftrightarrow f$ | $a \leftrightarrow g$ | $a \leftrightarrow g$ |
| $b \leftrightarrow g$ | $b \leftrightarrow e$ | $b \leftrightarrow f$ |
| $c \leftrightarrow e$ | $c \leftrightarrow f$ | $c \leftrightarrow e$ |

The number of pairings between two equivalent sets increases rapidly. For example, the natural pairing of a set of five coats with a set of five coat hangers would be to hang exactly one coat on each hanger. The number of different ways this can be done (different one-to-one correspondences) is one hundred twenty.

### Exercises I

Let $H = \{0, 1, 2, 3, 4, 5, 6, 7, 8, 9\}$, let $E = \{2, 4, 6, 8\}$, let $F = \{1, 3, 5, 7, 9\}$ and let $P = \{2, 3, 5, 7\}$. Use these sets to work Exercises 1 through 9.

1. Which of the following are true? (a) $7 \in H$. (b) $2 \in H$. (c) $2 \notin H$. (d) $2 \in F$. (e) $3 \notin P$. (f) $7 \in P$.
2. Which of the following are true? (a) $3 \in (F \cap P)$. (b) $3 \in (E \cap F)$. (c) $3 \in (E \cup F)$. (d) $3 \notin (P \cap H)$.
3. Which of the following are true? (a) $E \subseteq H$. (b) $P \subseteq H$. (c) $P \subseteq F$. (d) $\emptyset \subseteq F$.
4. List the elements in each of the following sets. (a) $E \cap H$. (b) $E \cup H$. (c) $E \cup F$. (d) $E \cap F$.
5. List the elements in each of the following sets. (a) $P \cup F$. (b) $E \cup P$. (c) $P \cap F$. (d) $E \cap P$.
6. In each of the following, the complement is relative to $H$. Identify the given set by listing the elements or by giving some other description of the set. (a) $E'$. (b) $F'$. (c) $P'$. (d) $H'$.
7. For the sets $H$, $E$, $F$, and $P$, list all pairs which are disjoint sets.
8. (a) What is $n(E)$? (b) What is $n(F)$? (c) What is $n(E \cup F)$? (d) What is $n(E \cap F)$?
9. (a) What is $n(F)$? (b) What is $n(P)$? (c) What is $n(F \cup P)$? (d) What is $n(F \cap P)$? (e) What is $n(E \cup \emptyset)$? (f) What is $n(E \cap \emptyset)$?
10. Are the sets $\{a, b, c\}$ and $\{e, f, g, h\}$ equivalent sets?

### Exercises II

Let $S = \{a, b, c, d, e, f, g, h\}$, let $A = \{a, b, c, d\}$, let $B = \{a, b, g, h\}$, and let $C = \{f, g, h\}$. Use these sets in Exercises 1 through 9 where all complements are relative to $S$.

1. List the elements in each of the following sets. (a) $A \cup B$. (b) $A \cap B$. (c) $A \cup S$. (d) $A \cap S$.
2. For the given sets, which pairs are disjoint sets?
3. Is $A \cup (B \cap C)$ the same set as $(A \cup B) \cap (A \cup C)$?
4. Is $A \cap (B \cup C)$ the same set as $(A \cap B) \cup (A \cap C)$?
5. Is $B \cup (A \cap C)$ the same set as $(B \cup A) \cap (B \cup C)$?

6. Is $B \cap (A \cup C)$ the same set as $(B \cap A) \cup (B \cap C)$?
7. List the elements in each of the following sets. (a) $A'$. (b) $B'$. (c) $C'$. (d) $(A \cup B \cup C)'$.
8. (a) Is $(A \cup B)'$ the same set as $A' \cap B'$? (b) Is $(A \cup C)'$ the same set as $A' \cap C'$? (c) Is $(B \cup C)'$ the same set as $B' \cap C'$?
9. (a) Is $(A \cap B)'$ the same set as $A' \cup B'$? (b) Is $(B \cap C)'$ the same set as $B' \cup C'$? (c) Is $(A \cap C)'$ the same set as $A' \cup C'$?
10. Let $S$ and $T$ be finite sets with $n(S)$ and $n(T)$ as cardinal numbers. (a) If $S \subseteq T$, what is $n(S \cup T)$? (b) If $S \subseteq T$, what is $n(S \cap T)$?
11. How many one-to-one correspondences are there between the equivalent sets $\{a, b, c, d\}$ and $\{e, f, g, h\}$?
12. List all the one-to-one correspondences between $\{a, b, c, d\}$ and $\{e, f, g, h\}$.

## 1.4 Supplementary Exercises

1. Name the position that "2" occupies in each of the following numbers. (a) 32. (b) 2,764. (c) 617,211. (d) 2,635,400. (e) 8,625,431.
2. Write each of the following in Hindu–Arabic notation and in Roman numeral notation. (a) Two hundred sixty-four. (b) Three thousand, seven hundred three.

State in words the name of each number given in Exercises 3 through 10.

3. 4,210.    4. 3,003.    5. 421,671.
6. XVIII.   7. LXXI.     8. CMXIV.

9. (a) The area of the United States is 3,540,939 square miles and the area of the Union of Soviet Socialist Republics is 8,646,489 square miles.
   (b) The area of Alaska is 566,432 square miles and the area of Rhode Island is 1,049 square miles.
10. (a) There are over 29,419,981 acres in the National Park System of the United States.
    (b) The area of China including Manchuria and Tibet is 3,691,502 square miles.
11. In the following, indicate in each case whether the number is used in the cardinal or ordinal sense.

(a) In the class of ten students, John is number one.
(b) Of the fifty states in the United States of America, California was the thirty-first state to be admitted into the Union.
(c) John has had three dogs since he was six years old.

12. Write each of the following numbers in Hindu–Arabic notation.
    (a) Eight thousand, four hundred twenty-six.
    (b) Two million, six hundred thirty.
    (c) Fifty-five million, six hundred thousand.
    (d) Two million, twenty thousand, two hundred.

13. Write each of the following numbers in Roman notation.
    (a) Nine hundred eighty-seven.
    (b) One thousand sixty-seven.
    (c) Two thousand, one hundred forty-two.
    (d) One thousand one.

14. Write each of the following in Hindu–Arabic notation.
    (a) MCML. (b) DLXVI. (c) MCMXLIV. (d) MDCLXVI.

In Exercises 15 through 19, use each of the following sets. $A = \{x, y, z\}$, $B = \{u, v, w\}$, $C = \{r, s, t, u, v\}$, $D = \{x, y, z, u, v, w\}$.

15. Which of the following are true? (a) $x \in A$. (b) $x \in C$. (c) $x \in (A \cup B)$. (d) $x \in (A \cap B)$. (e) $x \notin [D \cap (A \cup C)]$.

16. List the elements in each of the following sets. (a) $C \cap D$. (b) $B \cap D$. (c) $A \cap D$. (d) $D \cap (A \cup C)$.

17. Which of the following are true? (a) $A \subseteq D$. (b) $A'_D \subseteq D$. (c) $(A \cap B) \subseteq D$. (d) $B \subseteq C$. (e) $D \subseteq D$.

18. Show that each of the following is true.
    (a) $A \cap (C \cup D) = (A \cap C) \cup (A \cap D)$.
    (b) $B \cup (C \cap D) = (B \cup C) \cap (B \cup D)$.

19. For the sets $A$, $B$, $C$, and $D$, list all pairs of disjoint sets.

20. Let $S$ be the set of persons working for a company and define subsets of $S$ in the following way. (We assume that each subset is not empty.) Let $W$ be the set of women employed by the company, let $M$ be the set of men employed, let $U$ be the set of persons making less than $10,000, and let $V$ be the set of persons making $10,000 or more. Interpret each of the following statements. (a) $M \cup W = S$. (b) $W \subseteq U$. (c) $W \cap U = \emptyset$. (d) $W \cap V \neq \emptyset$. (e) $M \cap U = \emptyset$. (f) $M'_S \subseteq V$.

# 2

*abc*

$$\begin{array}{r}\mathbf{a}\\+\mathbf{b}\\\hline\mathbf{c}\end{array}\quad a+b=c$$

**+**

abcdefghijklmnopqrstuvwxyzabc

**=c**

# ADDITION OF WHOLE NUMBERS

## 2.1 Addition

After man began to concern himself with the number of elements in a set, it was inevitable that he would be concerned with the total number of elements in the union of two disjoint sets. For example, if one man has four sheep and his neighbor has nine sheep, how many do they both have?

Let $S = \{p, q, r\}$ and let $T = \{w, x, y, z\}$ be two *disjoint* sets. Then, by counting, we can determine that $n(S) = 3$, $n(T) = 4$ and $n(S \cup T) = 7$. This gives us a method to associate a whole number with the whole numbers 3 and 4. In general, if $a$ and $b$ are any two whole numbers, there exist sets $A$ and $B$ such that $A \cap B = \emptyset$, $n(A) = a$, and $n(B) = b$. If $c$ is the cardinal number of $A \cup B$, then $n(A \cup B) = c$ and we call $c$ the sum of $a$ and $b$. The sum of $a$ and $b$ is denoted by $a + b$; thus, $a + b = c$. This defines on the set of whole numbers the *operation* called addition; the symbol $+$ is used to denote this operation. Addition is called a binary operation since it assigns a unique whole number to any given pair of whole numbers. The numbers $a$ and $b$ which are added are called addends.

The statement $a + b = c$ is called an *equation*; it is a statement of equality. For example, the equation $3 + 4 = 2 + 5$ states that $3 + 4$ and $2 + 5$ are names for the same number. The equals sign ($=$) is used to denote logical identity. Although the symbols "2" and "II" are quite different, it is correct to write $2 = $ II since the equality pertains to the numbers and not to the notations used to denote the numbers.

As we have seen, it is convenient to use letters to denote numbers. For example, when we say "Let $a$ be a natural number," we mean that $a = 1$, $a = 2$, or that $a$ represents some other natural number. When we say "Let $a$ and $b$ be natural numbers," this does not preclude the fact that $a = b$; in other words, it might be that $a = 17$ and $b = 35$, or it might be that $a = 6$ and $b = 6$.

The following are the formalized statements of the familiar properties of equality.

1. If $a$ is any number, then $a = a$. This is called the reflexive property.
2. If $a$ and $b$ are numbers and if $a = b$, then $b = a$. This is called the symmetric property.
3. If $a$, $b$, and $c$ are numbers and if $a = b$ and $b = c$, then $a = c$. This is called the transitive property.

The symmetric and transitive properties can be used to show that "things equal to the same thing are equal to each other"; that is, if $a = b$ and $c = b$, then $b = c$ by the symmetric property, and from $a = b$ and $b = c$ we obtain $a = c$ from the transitive property. Although we make extensive use of these properties, it will usually be done in an informal way.

Let $a$ and $b$ be any whole numbers and let $A$ and $B$ be disjoint sets such that $n(A) = a$ and $n(B) = b$. Since $A \cup B = B \cup A$, we know that $n(A \cup B) = n(B \cup A)$. Therefore, we conclude that

$$a + b = b + a.$$

The commutative property of union for sets implies that addition is a commutative operation.

Consider the whole numbers 3, 4, and 6. The sets $S = \{f, g, h\}$, $T = \{p, q, r, s\}$, and $W = \{u, v, w, x, y, z\}$, for example, are three pairwise disjoint sets such that $n(S) = 3$, $n(T) = 4$, and $n(W) = 6$. When we say that the sets are *pairwise disjoint* we mean that $S \cap T = \emptyset$, $S \cap W = \emptyset$, and $T \cap W = \emptyset$. Since $n(S \cup T) = 7$ and $n(T \cup W) = 10$, it follows from the definition of addition that

$$3 + 4 = 7 \quad \text{and} \quad 4 + 6 = 10.$$

Now, $(S \cup T)$ and $W$ are disjoint sets; also, $S$ and $(T \cup W)$ are disjoint sets. By counting, we determine that $n((S \cup T) \cup W) = 13$ and $n(S \cup (T \cup W)) = 13$. Therefore,

$$(3 + 4) + 6 = 13 \quad \text{and} \quad 3 + (4 + 6) = 13.$$

In general, for any three whole numbers $a$, $b$, and $c$, there exist

## 2.1 Addition

pairwise disjoint sets $A$, $B$, and $C$ such that $n(A) = a$, $n(B) = b$, and $n(C) = c$. Since

$$A \cup (B \cup C) = (A \cup B) \cup C,$$

we conclude that

$$n(A \cup (B \cup C)) = n((A \cup B) \cup C).$$

Since $A$ and $(B \cup C)$ are disjoint sets and since $(A \cup B)$ and $C$ are also disjoint sets,

$$n(A) + n(B \cup C) = n(A \cup B) + n(C),$$

or

$$a + (b + c) = (a + b) + c.$$

We see that the associative property for the union of sets implies that addition is an associative operation.

To indicate the sum $3 + 4$, we often write

$$\begin{array}{rr} \text{Add:} & 3 \\ & \underline{4} \end{array}$$

Since $3 + 4 = 7$, we indicate this by

$$\begin{array}{rr} \text{Add:} & 3 \\ & \underline{4} \\ \text{Sum} & 7 \end{array}$$

Technically, we have added down in the vertical column of numbers. However, since $4 + 3 = 3 + 4$, we would also obtain the same sum by adding up the column of numbers. In fact, this is a technique often used to check answers in addition problems.

With the above agreement, we write

$$\begin{array}{rr} \text{Add:} & 3 \\ & 4 \\ & \underline{5} \end{array}$$

to represent the sum $(3 + 4) + 5$. The sum obtained by adding from the bottom of the vertical list would be $(5 + 4) + 3$. To prove that the sum would be the same we need to show that

$$(5 + 4) + 3 = (3 + 4) + 5.$$

This can be done by using the commutative and associative properties of addition in the following manner.

$(5 + 4) + 3 = (4 + 5) + 3$    Commutative property of addition.
$= 3 + (4 + 5)$    Commutative property of addition.
$= (3 + 4) + 5$    Associative property of addition.

Of course, we also use the familiar properties of equality in this proof.

Let $a$ be any whole number and let $A$ be a set such that $n(A) = a$. Since there are no elements in the empty set, it follows that $A$ and $\emptyset$ are disjoint sets. From the fact that

$$A \cup \emptyset = A,$$

it follows that

$$n(A \cup \emptyset) = n(A).$$

From the definition of addition, we have

$$n(A) + n(\emptyset) = n(A).$$

Since 0 is the cardinal number of the empty set, $n(\emptyset) = 0$ and

$$a + 0 = a.$$

An intuitively obvious fact is that 0 is the only number such that $a + 0 = a$. In other words, if $a$ and $b$ are any two whole numbers such that $a + b = a$, then $b = 0$.

Let us state the four basic properties of addition for the set of whole numbers. In Chapter 3, we give a more detailed discussion of Property 1.

**Property 1:** Closure property of addition. For any two whole numbers $a$ and $b$, the sum $a + b$ is a unique whole number.

**Property 2:** Commutative property of addition. For any pair of whole numbers $a$ and $b$, $a + b = b + a$.

**Property 3:** Associative property of addition. For any whole numbers $a$, $b$, and $c$, $a + (b + c) = (a + b) + c$.

**Property 4:** Additive identity. Let $a$ be any whole number. Then $a + 0 = a$ and 0 is called the *additive identity* for the set of whole numbers.

We can develop techniques for adding whole numbers which,

## 2.1 Addition

ADDITION TABLE

| + | 0 | 1 | 2 | 3 | 4 | 5 | 6 | 7 | 8 | 9 |
|---|---|---|---|---|---|---|---|---|---|---|
| 0 |   |   |   |   |   |   |   | 7 |   |   |
| 1 |   |   |   |   |   |   |   |   |   |   |
| 2 |   |   |   |   |   |   |   |   |   |   |
| 3 |   |   |   |   |   | 8 |   |   |   |   |
| 4 |   |   |   |   |   |   |   |   |   |   |
| 5 |   |   |   | 8 |   |   |   |   |   |   |
| 6 |   |   |   |   |   |   |   |   |   |   |
| 7 | 7 |   |   |   |   |   |   |   |   |   |
| 8 |   |   |   |   |   |   |   |   |   |   |
| 9 |   |   |   |   |   |   |   |   |   |   |

basically, require only that we know the sums of all possible pairs from $\{0, 1, 2, 3, 4, 5, 6, 7, 8, 9\}$. The number of sums of the form $a + b$ where $a$ and $b$ can each be any one of the ten given numbers is ten tens, or one hundred. Of course, since $3 + 5 = 5 + 3$, for example, the number of such sums which we need to learn is immediately reduced considerably. Generally, we make what is called an addition table in which to enter these number sums. The top row of the table is called the headline and the first column on the left is called the sideline. In both the headline and sideline, we list the numbers in the usual order. In the fourth row and sixth column we enter the sum $3 + 5$, where 3 is the fourth number in the sideline and 5 is the sixth number in the headline. Since $5 + 3 = 3 + 5$, we see that the sums will be symmetrically placed with respect to the line from the upper left-hand corner of the table to the lower right-hand corner; this line is called the main diagonal for the table.

### Exercises I

1. Complete the addition table on this page by supplying the missing entries.
2. Use the addition table to find the following sums.

(a) 6
    8̲

(b) 8
    8̲

(c) 9
    6̲

(d) 7
    9̲

(e) 5
    7̲

3. Use the addition properties and the entries in the addition table to find the following sums.

(a) 5
    4
    2
    4̲

(b) 6
    3
    4
    5̲

(c) 8
    1
    7
    2̲

(d) 6
    2
    4
    3̲

(e) 7
    2
    5
    3̲

4. In each case, determine what number $N$ represents. (a) $9 + 6 = N$. (b) $8 + N = 8$. (c) $0 + N = 0$. (d) $7 + 9 = N$. (e) $9 + N = 17$. (f) $5 + N = 13$.

5. (a) In the addition table, what can be said about the row of entries where the additive identity appears in the sideline?
   (b) In the addition table, what can be said about the column of entries where the additive identity appears in the headline?
   (c) Starting at the sideline, explain how counting one at a time can be used to fill in the table.

6. Let $A = \{a, b, c, d\}$ and let $B = \{c, d, e, f\}$. (a) What is $n(A)$? (b) What is $n(B)$? (c) What is $n(A \cup B)$? (d) Is $n(A) + n(B) = n(A \cup B)$?

7. Let $S = \{a, b, c, d, e, f\}$ and let $T = \{c, d, e, g, h\}$. (a) What is $n(S)$? (b) What is $n(T)$? (c) What is $n(S \cup T)$? (d) Is $n(S) + n(T) = n(S \cup T)$?

8. (a) In Exercise 6, what is $n(A \cap B)$? (b) In Exercise 7, what is $n(S \cap T)$?

9. (a) In Exercise 6, is $n(A \cup B) + n(A \cap B) = n(A) + n(B)$?
   (b) In Exercise 7, is $n(S \cup T) + n(S \cap T) = n(S) + n(T)$?

10. Let $A$ and $B$ be two finite sets where $n(A) = a$ and $n(B) = b$. Give an argument to show that

$$n(A \cup B) + n(A \cap B) = n(A) + n(B).$$

*Hint*: See Exercises 6, 7, 8, and 9.

## Exercises II

1. Let $a$, $b$, $c$, and $d$ be whole numbers and consider the following.

## 2.1 Addition

$$\text{Add:} \quad \begin{array}{c} a \\ b \\ c \\ \underline{d} \end{array}$$

The sum obtained by adding from top to bottom is $((a + b) + c) + d$ and the sum obtained from adding from bottom to top is $((d + c) + b) + a$. Use the commutative and associative properties of addition and properties of equality to prove that the two sums are the same.

In Exercises 2 through 6, use the result in Exercise 10 above.

2. In a class of 30 students, 27 passed the first test and 24 passed the second. No one failed both tests. How many students passed both tests?

3. In a class of 73 students, 68 passed the first test and 65 passed the second. No one failed both tests. How many students passed both tests?

4. In two training sessions for 60 salesmen, 52 attended the first session and 49 attended the second. No salesman missed both sessions. How many salesmen attended both sessions?

5. In a group of 1,000 car owners who owned Makes $A$ or $B$, it was found that 627 owned Make $A$ and 831 owned Make $B$. How many of the group owned both Makes $A$ and $B$?

6. (a) In a questionnaire to 500 persons who drank tea or coffee, it was found that 376 drank tea and 412 drank coffee. How many drank both tea and coffee?
   (b) How many drank tea only?
   (c) How many drank coffee only?

*7. If $P$, $Q$, and $R$ are sets, is it true that
$$n((P \cap Q) \cap (Q \cap R)) = n(P \cap Q \cap R)?$$

*8. Let $P$, $Q$, and $R$ be sets. Use the tenth exercise in Exercises I to conclude that
$$n(P \cap (Q \cup R)) + n(P \cup (Q \cup R)) = n(P) + n(Q \cup R).$$

*9. For sets $P$, $Q$, and $R$, is it true that
$$n(P \cap (Q \cup R)) = n((P \cap Q) \cup (P \cap R))?$$

*10. Use Exercises 7, 8, and 9 to verify each of the following statements for sets $P$, $Q$, and $R$.

(1) $n(P \cap (Q \cup R)) + n(P \cup (Q \cup R)) = n(P) + n(Q \cup R)$
(2) $n(P \cap (Q \cup R)) + n(P \cup Q \cup R) + n(Q \cap R) = n(P) + n(Q) + n(R)$
(3) $n((P \cap Q) \cup (P \cap R)) + n(P \cup Q \cup R) + n(Q \cap R) = n(P) + n(Q) + n(R)$
(4) $n(P \cap Q) + n(P \cap R) + n(P \cup Q \cup R) + n(Q \cap R) = n(P) + n(Q) + n(R) + n((P \cap Q) \cap (Q \cap R))$
(5) $n(P \cup Q \cup R) + n(P \cap Q) + n(P \cap R) + n(Q \cap R) = n(P) + n(Q) + n(R) + n(P \cap Q \cap R)$

## 2.2 Techniques for Addition

Suppose we wish to find the sum $37 + 48$. Since

37 is 3 tens and 7 units

and

48 is 4 tens and 8 units,

it follows that

$37 + 48$ is $(3 + 4)$ tens and $(7 + 8)$ units.

Since $7 + 8 = 15$, or 1 ten and 5 units,

$37 + 48$ is $(3 + 4 + 1)$ tens and 5 units.

Thus,

$37 + 48 = 85.$

Using the fact that $37 = 30 + 7$ and $48 = 40 + 8$, we see that

$37 + 48 = (30 + 7) + (40 + 8).$

Using the associative and commutative properties of addition,

$37 + 48 = (30 + 40) + (7 + 8)$
$= 70 + 15$
$= 85.$

We can perform this addition by writing the numbers in a vertical column as follows.

## 2.2 Techniques for Addition

$$\begin{array}{r} 37 \\ 48 \\ \underline{15} \\ 70 \\ \underline{85} \end{array}$$

We add the numbers in the units' position, add the numbers in the tens' position, and place the sums in the vertical column with units' and tens' positions aligned. Then, we add the two resulting sums to obtain the sum 85.

Another and somewhat shorter procedure to perform the addition uses what is called the "carrying" process. Instead of writing down the sum 15, we write "5" in the units' position and "carry" 1 to be added in the tens' column. Then, we find the sum $(1 + 3 + 4)$. This technique is indicated as follows.

$$\begin{array}{r} 1 \\ 37 \\ \underline{48} \\ 85 \end{array}$$

If we are to find the sum of several numbers, then we may use various techniques. Let us exhibit and discuss three of them.

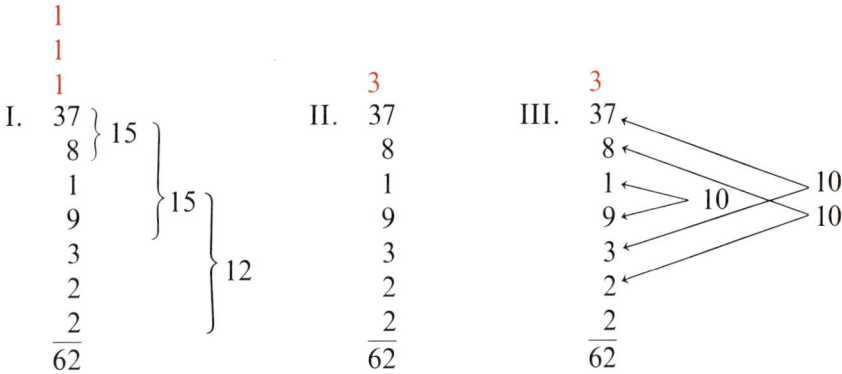

**Method I.** We add 7 and 8 and obtain the sum 15. "Carry" 1 to the second column and retain the 5. Then, find the sum $5 + 1 + 9$; this is 15. Retain the 5 and carry another 1 to the tens' column. Since $5 + 3 + 2 + 2 = 12$, write 2 in the units' column and carry another 1 to the tens' column. Adding $1 + 1 + 3$ in the tens' column, we obtain 6 and then conclude that the sum is 62.

30                                                              *Addition of Whole Numbers*

**Method II.** The technique here differs only slightly from Method I. After gaining a little experience, we add the numbers in the units' column and obtain a sum of 32. We write 2 in the units' column, carry 3 to the tens' column, and find the sum of the numbers in the tens' column.

**Method III.** Here, we take more than a casual notice of the fact that addition is a commutative and an associative operation and add the numbers in the units' column in a convenient order. We seek familiar combinations, such as $7 + 3, 8 + 2$, and $1 + 9$. Since each pair chosen has 10 as the sum, these six numbers have 30 as their sum. Since 2 is the only number which has not been added, the sum of the numbers in the first column is 32. Now, find the sum of the numbers in the tens column as before.

*Examples*

```
                                                    43
 1.   268                              2.   268
      385                                   385
      462                                   462
      581                                   581
      487                                   487
      268                                   268
     ────                                  ────
       31   (units' column)                2451
      420   (tens' column)
     2000   (hundreds' column)
     ────
     2451
```

3.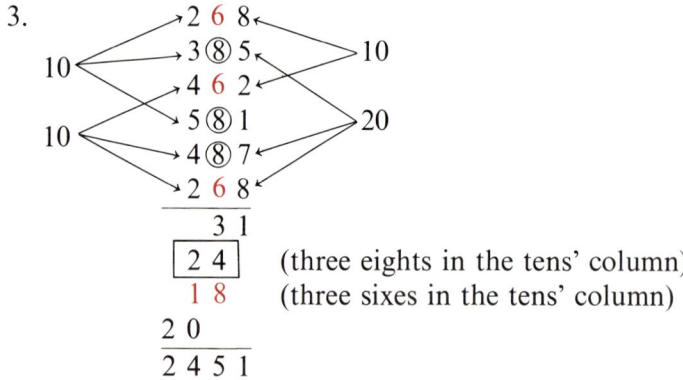

(three eights in the tens' column)
(three sixes in the tens' column)

## 2.2 Techniques for Addition

We should clearly distinguish between what the sum of several numbers is and the technique used to find the sum. As we have seen, the sum of a given set of numbers is unique but the method used to find the sum is not. After considerable practice, many persons add two, or more, columns of numbers at a time. Within certain limits, the facility with which one adds a column of numbers varies with the amount of concentrated practice.

We can also find the sum of a column by numbers by adding successively two numbers at a time. This technique is exhibited as follows.

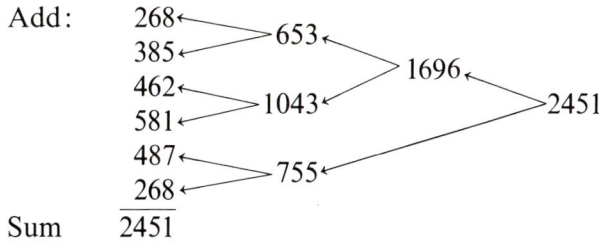

Of course, another technique is to use a calculating machine. At present, this technique should be avoided since it would defeat our purposes and impede attaining the goals of the text.

### Exercises I

In Exercises 1 through 10, find the indicated sums. Do not use the same technique for all problems.

1. (a) 38　　(b) 97　　(c) 75
　　　25　　　　35　　　　57

2. (a) 54　　(b) 93　　(c) 35
　　　32　　　　45　　　　16
　　　66　　　　28　　　　85

3. (a) 847　(b) 598　(c) 576
　　　488　　　379　　　123

4. (a) 378　(b) 863　(c) 722
　　　544　　　296　　　466
　　　299　　　127　　　228

5. (a) 65　　(b) 54　　(c) 18
　　　87　　　　 33　　　　 45
　　　11　　　　 66　　　　 75
　　　32　　　　 87　　　　 92

6. (a) 1,476　(b) 3,896　(c) 4,771
　　　3,799　　　7,223　　　3,448

7. (a) 2,478　(b) 9,887　(c) 4,665
　　　6,889　　　3,618　　　9,143
　　　4,552　　　1,546　　　3,333

8. (a) 24,576　(b) 10,261　(c) 54,881
　　　38,900　　　45,872　　　35,133
　　　31,233　　　44,918　　　42,678

9. (a) 3,891　(b) 3,899　(c) 8,882
　　　5,904　　　4,353　　　1,655
　　　3,567　　　1,037　　　3,258
　　　4,321　　　2,121　　　2,015

10. (a) 45,213　(b) 31,555　(c) 34,888
　　　32,588　　　12,003　　　21,012
　　　11,001　　　22,542　　　44,442
　　　34,109　　　13,578　　　15,022
　　　24,112　　　34,645　　　88,989

## Exercises II

Find the indicated sums in Exercises 1 through 4.

1. 6,284　　2. 8,425　　3. 4,566　　4. 4,567
　 1,481　　　 9,101　　　 3,040　　　 2,899
　 2,672　　　 2,200　　　 2,114　　　 3,141
　 8,231　　　 3,215　　　 3,333　　　 3,444

5. Find the sum of the four top numbers in each of the columns of numbers in Exercises 1 through 4.
6. Find the sum of the four second numbers in each of the columns of numbers in Exercises 1 through 4.
7. Find the sum of the four third numbers in each of the columns of numbers in Exercises 1 through 4.

8. Find the sum of the four fourth numbers in each of the columns of numbers in Exercises 1 through 4.
9. Find the sum of the four sums found in Exercises 1 through 4.
10. (a) Find the sum of the four sums found in Exercises 5 through 8.
    (b) Does the answer in part (a) check with the answer in Exercise 9?

## 2.3 Checks for Addition

We found in the last section that one way to check an addition problem is to find the sum by more than one method. This technique to check a sum, as well as most checks for the arithmetic operations, involves some other computational procedures. Therefore, it is possible, though unlikely, that an incorrect answer may "check" as a result of making two or more computational errors. For this reason, most checks should be called either *pseudo-checks* or some name which indicates that the possibility of error has not been completely eliminated. If, however, we are aware of this shortcoming, then using the standard terminology and identifying any such procedure as a "check" should cause little difficulty.

Let us discuss one more check for addition. This check is quite different from the ones indicated in Section 2.2 since we do not just find the sum of a column of numbers in more than one way. Consider the following.

$$
\begin{array}{lllllll}
\text{Add:} & 368 & \xrightarrow{+} 17 & \xrightarrow{+} & & \xrightarrow{+} & 8 \\
 & 429 & \xrightarrow{+} 15 & \xrightarrow{+} & & \xrightarrow{+} & 6 \\
 & 766 & \xrightarrow{+} 19 & \xrightarrow{+} 10 & \xrightarrow{+} & 1 \\
 & 321 & & \xrightarrow{+} & & \xrightarrow{} & 6 \\
\hline
\text{Sum} & 1884 & \xrightarrow{+} 21 & \xrightarrow{+} 3 & & 21 & \xrightarrow{+} 3
\end{array}
$$

*Explanation of the Procedure.* Add the digits in each addend and record the sum at the right. Repeat the process for each of the resulting sums until a one-digit number is obtained. Then, add the one-digit numbers; in the example, the sum is 21. If the resulting sum is not a one-digit number, then add the digits in the number. Repeat the process until a one-digit number is obtained; in the example, 2 + 1 is 3. Now, add the digits in the original sum and continue the process until a one-digit number is obtained; in the example, 2 + 1 is 3. If the two final one-digit numbers are different, then an error has been made.

The check just discussed is sometimes called *casting-out-nines*. We shall have more to say about this check later; at that time we shall give the reason for choosing the name "casting-out-nines" for the checking procedure. (Hopefully, the student will discover the answer before then.)

### Examples

1.  $872 \xrightarrow{+} 17 \xrightarrow{+} 8$
    $384 \xrightarrow{+} 15 \xrightarrow{+} 6$
    $321 \xrightarrow{\phantom{+}+\phantom{+}} 6$
    $437 \xrightarrow{+} 14 \xrightarrow{+} 5$
    $\overline{2014} \xrightarrow{+} 7 \qquad 25 \xrightarrow{+} 7$

2.  $436 \xrightarrow{+} 13 \xrightarrow{+} 4$
    $271 \xrightarrow{+} 10 \xrightarrow{+} 1$
    $483 \xrightarrow{+} 15 \xrightarrow{+} 6$
    $956 \xrightarrow{+} 20 \xrightarrow{+} 2$
    $\overline{2146} \xrightarrow{+} 13 \xrightarrow{+} 4 \quad 13 \xrightarrow{+} 4$

For the casting-out-nines check, let us emphasize that we did not say if the final one-digit numbers are the same then the sum is correct. In general, this implication is not true. For example, in the first addition problem if the incorrect answer 1924 were obtained, then

$$1924 \xrightarrow{+} 16 \xrightarrow{+} 7,$$

and the final one-digit numbers would agree. However, this type of error is rather unlikely and casting-out-nines is usually a reliable check. Of course, as previously stated, if the final one-digit numbers do not agree we know that some error has been made.

### Exercises I

Using the casting-out-nines technique, check the addition problems in each of the indicated exercises of Exercises I, Section 2.2.

## 2.3 Checks for Addition

1. Exercise 7a.
2. Exercise 7b.
3. Exercise 7c.
4. Exercise 8a.
5. Exercise 8b.
6. Exercise 8c.
7. Exercise 9a.
8. Exercise 9b.
9. Exercise 9c.
10. Exercise 10a.
11. Exercise 10b.
12. Exercise 10c.

## 2.4 Supplementary Exercises

In each of the Exercises 1 through 9, determine what number $N$ represents.

1. $8 + N = 15$.
2. $7 + 9 = N$.
3. $6 + N = 6$.
4. $7 + N = 16$.
5. $17 + 0 = N$.
6. $23 + 72 = N$.
7. $16 + N = 7 + 16$.
8. $(8 + N) + 3 = 15$.
9. $(2 + N) + 13 = 19$.

Let $A = \{a, b, c, d, e, f\}$, let $B = \{c, e, f, g, h\}$, and let $C = \{e, f, m, n, p, q, r, s, t\}$. Use these sets in Exercises 10 through 14.

10. (a) What is $n(A)$? (b) What is $n(B)$? (c) What is $n(C)$?
11. (a) What is $n(A \cup B)$? (b) What is $n(A \cup C)$? (c) What is $n(B \cup C)$? (d) What is $n(A \cup B \cup C)$?
12. (a) What is $n(A \cap B)$? (b) What is $n(A \cap C)$? (c) What is $n(B \cap C)$? (d) What is $n(A \cap B \cap C)$?
13. (a) Show that $n(A \cup B) + n(A \cap B) = n(A) + n(B)$.
    (b) Show that $n(A \cup C) + n(A \cap C) = n(A) + n(C)$.
    (c) Show that $n(B \cup C) + n(B \cap C) = n(B) + n(C)$.
14. Show that $n(A \cup B \cup C) + n(A \cap B) + n(A \cap C) + n(B \cap C) = n(A) + n(B) + n(C) + n(A \cap B \cap C)$.

In each of Exercises 15 through 20, find the sum and check by the casting-out-nines technique.

15. 23,576
    41,784
    38,558
    51,221
    32,569

16. 45,890
    12,211
    40,548
    72,555
    87,340

17. 19,830
    32,477
    58,947
    11,856
    78,324

18. 132,576
    357,989
    430,870
    358,749
    567,987

19. 458,977
    200,050
    114,671
    456,876
    318,592

20. 352,222
    302,001
    878,982
    443,890
    123,456

**3**

$$\frac{\begin{array}{r}c\\-d\end{array}}{e}$$

$a - b = c$
$c - d = e$
$a > c$
$d < c$

$$\frac{\begin{array}{r}a\\-b\end{array}}{c}$$

$a - a = 0 \quad a < b$
$c > d \quad b - d = e$

# SUBTRACTION OF WHOLE NUMBERS

## 3.1 Subtraction

The operation of subtraction evolves quite naturally from the operation of addition. Suppose a man has six sheep and his neighbor has fifteen. It is natural to want to know how many more sheep the neighbor has. In other words, what is the number $N$ such that the sum of 6 and $N$ is 15?

The number $N$ such that $6 + N = 15$ is called the *difference* of 15 subtract 6. In general, for any numbers $a$ and $b$, if there exists a unique number $c$ such that $a + c = b$ then $c$ is called the difference of $b$ subtract $a$. Symbolically,

$$b - a = c \quad \text{if and only if} \quad a + c = b.$$

For the difference $b - a$, the number $b$ is called the minuend and $a$ is called the subtrahend.

Subtraction presents one difficulty not encountered with addition. There is no whole number $N$, for example, such that $12 + N = 8$; that is, there is no whole number $N$ which is the difference of 8 subtract 12. For a binary operation, such as addition,

> if the operation always assigns a unique number in the set to any given pair in the set, then the set is said to be closed with respect to the operation.

For example, since the *sum* of two whole numbers is always a whole number, the set of whole numbers is closed with respect to addition. However, the set of whole numbers is not closed with respect to subtraction. The fact that the set of whole numbers does not have the closure property with respect to subtraction is one reason why we eventually enlarge our number system to include the negative numbers. But, at present, the lack of closure for subtraction is a difficulty we shall have to endure.

The definition of subtraction makes the following problem sets equivalent.

Find the missing number in each of the following.

A. Add:

| | | | | | |
|---|---|---|---|---|---|
| | 6 | 8 | 17 | 11 | 13 |
| | $N$ | $N$ | $N$ | $N$ | $N$ |
| Sum | 15 | 14 | 25 | 18 | 18 |

B. Subtract:

| | | | | | |
|---|---|---|---|---|---|
| | 15 | 14 | 25 | 18 | 18 |
| | 6 | 8 | 17 | 11 | 13 |
| Difference | $N$ | $N$ | $N$ | $N$ | $N$ |

From our first view of subtraction, finding the difference $15 - 6$ is equivalent to finding *how many more* objects are in a set with 15 objects than in one with 6 objects. Another way to obtain the difference is to determine how many objects would be left after taking away 6 objects from a set with 15 objects.

The two points of view concerning subtraction are reflected in our techniques for performing the operation. For example, in the following problem

Subtract: 13
7
Difference 6

we can ask either one of the following questions to obtain the difference. (1) What do we add to 7 to get a sum of 13; that is, how many objects do we put into a set with 7 objects in order to have a set with 13 objects? (2) If we take away 7 objects from a set containing 13 objects, how many objects are left in the original set? Both of these techniques are used to find differences; the better one to choose is probably the one which seems to be easier (or more familiar).

Let $a$ be any whole number. Since $a + 0 = a$, it follows from the definition of subtraction that

$$a - a = 0.$$

Similarly, since $0 + a = a$, it follows that

$$a - 0 = a.$$

## 3.1 Subtraction

It should be immediately obvious that subtraction is *not* a commutative operation. Even when the differences exist, we can show that subtraction is not an associative operation either. Consider the differences

$$(15 - 10) - 2 \quad \text{and} \quad 15 - (10 - 2).$$
$$(15 - 10) - 2 = 5 - 2 = 3 \quad \text{and} \quad 15 - (10 - 2) = 15 - 8 = 7.$$

Thus,

$$(15 - 10) - 2 \neq 15 - (10 - 2).$$

Sometimes, subtraction is called the "inverse" operation of addition. When this terminology is used one generally means to point out that subtracting a number which was previously added "undoes" the effect of the original addition.

### Exercises I

1. Use the addition table on page 25 to find the following differences.
   Subtract: (a) 17   (b) 16   (c) 12   (d) 13
              9        7        7        6

2. Use the addition table on page 25 to find each of the following differences. (a) $17 - (8 - 3)$. (b) $(17 - 8) - 3$. (c) $18 - (9 - 4)$. (d) $(18 - 9) - 4$.

3. (a) If a subtraction table were made similar to the addition table on page 25, what number would appear on the main diagonal?
   (b) Using only whole numbers and the standard procedure for listing the entries in the table, which side of the main diagonal would not contain any entries in a subtraction table?

Perform the indicated operations in Exercises 4 through 10.

4. $((18 - 9) + 4) - 6$.
5. $((13 - 7) + 2) - 4$.
6. $((14 - 0) - 7) + (8 - 5)$.
7. $((17 - 8) + 6) - ((6 + 2) - 3)$.
8. $((15 + 2) - 9) + ((11 - 3) - 2)$.
9. $((13 - 9) - 4) + (12 - (8 + 4))$.
10. $[(14 - 7) + (11 - 3) + (6 - 2)] - (15 - 8)$.

### Exercises II

In Exercises 1 through 6, use the following sets. Let $B = \{p, r, s, t, u, v, w, x, y, z\}$ and $A = \{r, s, t, u, v, w\}$.
1. (a) What is $n(A)$? (b) What is $n(B)$?
2. (a) Is $A \subseteq B$? (b) Is $B \subseteq A$?
3. (a) What is $A \cup B$? (b) What is $A \cap B$?
4. (a) What is $A'$, the complement of $A$ relative to $B$?
   (b) What is $n(A')$?
5. (a) Is $A \cup A' = B$? (b) Is $A \cap A' = \emptyset$?
6. Using the fact that $n(A \cup A') + n(A \cap A') = n(A) + n(A')$, conclude that $n(B) - n(A) = n(A')$.
7. Let $S$ and $T$ be any finite sets such that $S \subseteq T$. Show that $n(T) - n(S) = n(S'_T)$.
8. Is $(12 - 4) + (7 - 3) = (12 + 7) - (4 + 3)$?
9. Is $(15 - 9) + (8 - 5) = (15 + 8) - (9 + 5)$?
10. Let $a$, $b$, $d$, and $e$ be whole numbers such that $a - b$ and $d - e$ are the whole numbers $c$ and $f$, respectively. Prove that $(a - b) + (d - e) = (a + d) - (b + e)$.

## 3.2 Techniques for Subtraction

To find the difference $59 - 25$, we could proceed as follows. Since

$$59 \text{ is } 5 \text{ tens and } 9 \text{ units}$$

and

$$25 \text{ is } 2 \text{ tens and } 5 \text{ units,}$$

we see that 59 is 3 tens and 4 units more than 25; that is,

$$59 - 25 = 34.$$

We may also perform the subtraction by columns as we did for addition.

$$\begin{array}{rr} \text{Subtract:} & 59 \\ & 25 \\ \hline \text{Difference} & 34 \end{array}$$

There is one difficulty encountered with the column technique for subtraction. It is exhibited in the following problem.

## 3.2 Techniques for Subtraction

$$\text{Subtract:} \quad \begin{array}{r} 64 \\ 26 \\ \hline \end{array}$$

There is no whole number we can add to 6 to get 4; therefore, we cannot just subtract the numbers in the units' position. Here, we make use of the fact that

$$\begin{aligned} 64 &= 6 \text{ tens} + 4 \text{ units} \\ &= (5 \text{ tens} + 1 \text{ ten}) + 4 \text{ units} \\ &= 5 \text{ tens} + (1 \text{ ten} + 4 \text{ units}) \\ &= 5 \text{ tens} + 14 \text{ units.} \end{aligned}$$

Since 14 is 8 more than 6 and 5 is 3 more than 2,

$$64 - 26 = 38.$$

The usual procedure is to use a shorter method to rewrite 64 as 5 tens plus 14 units; it is called *borrowing* and is exhibited as follows.

$$\begin{array}{r} 5 \\ 6^1 4 \\ 2\ 6 \\ \hline 3\ 8 \end{array}$$

We "borrow" 1 ten from the tens' column which reduces 6 tens to 5 tens and increases 4 to 14. Then, we find the differences by columns. This procedure generalizes for use in each of the other columns.

### Examples

1. $\begin{array}{r} 8\ 4 \\ 2\ 3 \\ \hline 6\ 1 \end{array}$ 2. $\begin{array}{r} 4 \\ \cancel{5}^1 3 \\ 2\ 7 \\ \hline 2\ 6 \end{array}$ 3. $\begin{array}{r} 1 \\ 6\ 2^1 6 \\ 4\ 0\ 9 \\ \hline 2\ 1\ 7 \end{array}$ 4. $\begin{array}{r} 7^1 3 \\ 8\ 4^1 5 \\ 2\ 7\ 8 \\ \hline 5\ 6\ 7 \end{array}$

There is one obvious method to check a subtraction problem such as the following.

$$\text{Subtract:} \quad \begin{array}{ll} 23{,}487 & \text{(minuend)} \\ 12{,}266 & \text{(subtrahend)} \\ \hline 11{,}221 & \text{(difference)} \end{array}$$

We find the sum of the difference and the subtrahend.

$$\begin{array}{ll} \text{Add:} & 11{,}221 \\ & 12{,}266 \\ \hline \text{Sum} & 23{,}487 \end{array}$$

If the sum is the same as the minuend in the subtraction problem, then the answer is correct.

Another method of check involves the casting-out-nines check discussed for addition. It is exhibited as follows.

$$\begin{array}{rl} \text{Subtract:} & 23{,}487 \xrightarrow{+} 24 \xrightarrow{+} 6 \\ & 12{,}266 \xrightarrow{+} 17 \xrightarrow{+} 8 \\ \text{Difference} & 11{,}221 \xrightarrow{+} 7 \\ & \text{Sum} \quad 15 \xrightarrow{+} 6 \end{array}$$

We should observe that the casting-out-nines check for subtraction is used to show that the *sum* of the difference and subtrahend is the minuend.

## Exercises I

In Exercises 1 through 12, find the differences and check your answers.

1. 478
   232

2. 687
   423

3. 835
   418

4. 937
   518

5. 723
   378

6. 782
   496

7. 3,748
   1,688

8. 4,822
   2,765

9. 43,674
   28,849

10. 673,425
    287,878

11. 467,900
    158,923

12. 478,902
    381,088

## Exercises II

In Exercises 1 through 12, find the differences and check your answers by the casting-out-nines check.

1. 2,861
   421

2. 8,241
   3,128

3. 68,251
   23,000

## 3.3 Less Than and Greater Than

| | | |
|---|---|---|
| 4. 31,482<br>   23,846 | 5. 80,007<br>   31,458 | 6. 103,264<br>    88,421 |
| 7. 204,361<br>    78,942 | 8. 586,481<br>   432,847 | 9. 3,682,425<br>   1,691,537 |
| 10. 83,254,512<br>    37,305,735 | 11. 43,566,870<br>    21,823,445 | 12. 77,886,923<br>    38,228,296 |

## 3.3 Less Than and Greater Than

Let $a$ and $b$ be any two whole numbers. If there exists a **natural number** $x$ such that $a + x = b$, then $a$ is said to be **less than** $b$; this is symbolized by $a < b$. For example, $3 < 7$ since there is a natural number $x$, namely 4, such that $3 + x = 7$. Of course, another way to show that $a < b$ is to show that the difference $b - a$ is a natural number.

### Examples

1. $0 < 6$.  2. $3 < 11$.  3. $5 < 10$.  4. $1 < 100$.

Let $a$ and $b$ be whole numbers. The number $a$ is said to be **greater than** $b$ if and only if $b$ is less than $a$. We use $b > a$ to state that $b$ is greater than $a$. Thus,

$$b > a \quad \text{if and only if} \quad a < b.$$

We call the symbol $<$ the *less-than* symbol, we call the symbol $>$ the *greater-than* symbol, and we call the statements $a < b$ and $b > a$ *inequalities*.

We sometimes write $a \leq b$ to indicate that $a < b$ or $a = b$. For example, if $x$ is a whole number such that $x \leq 3$ then $x$ is one of the numbers 0, 1, 2, or 3. Similarly, we write $a \geq b$ to indicate that $a > b$ or $a = b$. We should note that we can conclude from $c \leq d$ and $c \geq d$ that $c = d$.

If $a$ and $b$ are whole numbers, then one and only one of the following is true: $a < b$, $a = b$, or $a > b$; this is called the **trichotomy property**. The trichotomy property is equivalent to stating for finite sets $A$ and $B$, where $n(A) = a$ and $n(B) = b$, that set $A$ is smaller than set $B$, set $A$ is the same size as set $B$, or set $A$ is larger than set $B$.

For whole numbers $a$, $b$, and $c$, if $a < b$ and $b < c$ then $a < c$; this is called the transitive property of the less-than relation. The transitive property is equivalent to stating for finite sets $A$, $B$, $C$ that if $A$ is smaller than $B$ and $B$ is smaller than $C$, then $A$ is smaller than $C$.

Both the trichotomy and transitive properties of the less-than relation are important. Any set which has both the trichotomy and and transitive properties is called an ordered set. We make rather extensive use of the ordered set properties in the development of arithmetic.

Let $S$ be a set of numbers. If there exists a number $a$ in $S$ such that $a \leq x$ for *every* $x$ in $S$, then $a$ is called the least element in $S$. Similarly, if there exists a number $b$ in $S$ such that $b \geq x$ for *every* $x$ in $S$, then $b$ is called the greatest element in $S$. For example, if $S = \{2, 6, 11, 14\}$, then 2 is the least element in $S$ and 14 is the greatest element in $S$.

Two important facts concerning greatest and least elements for sets of whole numbers are the following. The set of whole numbers does *not* contain a greatest element; in other words, there is no last counting number. However, it is true that *any* (nonempty) set of whole numbers contains a least element; this is called the well-ordering property for the set of whole numbers. "Well-ordering" is a rather impressive name for a rather simple property, but it is important to remember since we shall make use of it later.

## Exercises I

1. Which of the following are true? (a) $3 < 6$, (b) $0 < 7$, (c) $5 > 9$, (d) $17 \leq 17$, (e) $3 \leq 5$, (f) $6 < 6$.
2. Which of the following are true? (a) $1 < 0$, (b) $88 < 91$, (c) $29 \geq 13$, (d) $0 \geq 0$, (e) $13 \leq 17$, (f) $17 > 17$.

In Exercises 3 through 10, determine the least element in each set, and determine the greatest element if the set has one.

3. $\{3, 5, 8, 11, 13, 23\}$.
4. $\{1, 3, 5, 7, 9, 11, 13\}$.
5. $\{x$ is a whole number and $x < 8\}$.
6. $\{x$ is a natural number and $x < 11\}$.
7. $\{x$ is a whole number and $x > 14\}$.
8. $\{x$ is a whole number and $x \geq 23\}$.

9. {$x$ is a whole number, $x \geq 3$, and $x < 33$}.
10. {$x$ is a whole number and $x + 3 < 10$}.

## Exercises II

In Exercises 1 through 4, determine the least element in the set, and determine the greatest element if the set has one.

1. {$x$ is a natural number and $x < 17$}.
2. {$x$ is a natural number, $x \leq 19$, and $x > 3$}.
3. {$x$ is a natural number, $x < 32$, and $x \geq 5$}.
4. {$x$ is a natural number, $x + 4 > 7$, and $x - 3 < 18$}.
5. Is it true for whole numbers $a$, $b$, and $c$ that if $a > b$ and $b > c$ then $a > c$?
6. Use the definition of less-than and the properties of addition of whole numbers to prove the transitive property of less-than.
7. Let $a$, $b$, and $c$ be whole numbers. Prove if $a < b$ then $a + c < b + c$.
8. Let $a$, $b$, and $c$ be whole numbers. Prove if $a \leq b$ and $b < c$ then $a < c$.
9. Let $a$, $b$, $c$, and $d$ be whole numbers. Prove if $a < b$ and $c < d$ then $a + c < b + d$.
10. Let $a$, $b$, and $c$ be whole numbers. Prove if $a + c < b + c$ then $a < b$.

## 3.4 Supplementary Exercises

1. Which of the following are true? (a) $6 \geq 6$. (b) $7 \leq 7$. (c) $0 < 6$. (d) $1 > 2$. (e) $4 \leq 7$. (f) $17 > 5$.
2. Let $a$, $b$, $c$, and $d$ be whole numbers such that $a < b$ and $c < d$. If $a - c$ and $b - d$ are whole numbers, can we prove that $a - c < b - d$?

For each of the sets in Exercises 3 through 6, state what the least element in the set is and state what the greatest element is if the set has one.

3. {$x$ is a natural number and $x < 15$}.
4. {$x$ is a whole number and $x \leq 22$}.

5. {$x$ is a whole number and $x > 35$}.
6. {$x$ is a whole number and $x + 4 \geq 28$}.
7. Is $[(18 - 3) + (15 - 2)] - 8 < [(17 - 2) - 3] + 10$?
8. Is $[(28 - 17) - 11] + 6 < [(14 + 3) - 15] + 4$?
9. If $(36 - N) + 5 = 23 - 17$, what is $N$?
10. If $(N + 25) - 8 = 31 + 92$, what is $N$?
11. If $(82 - N) + 17 = N + 23$, what is $N$?

In each of Exercises 12 through 20, find the difference and check your answer by the casting-out-nines check.

12. 23,578
    11,897

13. 45,822
    21,577

14. 60,002
    23,218

15. 123,589
    114,729

16. 234,567
    233,898

17. 654,321
    123,456

18. 570,334
    243,800

19. 932,421
    285,768

20. 480,020
    109,938

**4**

$6 \times 3 = 3 \times 6 = 3 + 3 + 3 + 3 + 3 + 3 = 6 + 6 + 6$

\* \* \* \* \*
\* \* \* \* \*
\* \* \* \* \*

# XVIII

**a × b**
**a · b**
**a × 0 = 0 × a = 0**

# MULTIPLICATION OF WHOLE NUMBERS

## 4.1 Multiplication

The underlying feature of the operation of multiplication for whole numbers is the grouping of elements into sets in which each has the same cardinal number. Suppose we wanted to know how many dots there are in the set pictured in Figure 1. Instead of counting the dots one at a time, we might observe that there are 6 subsets with 3 dots in each subset. The total number is the sum of 6 threes, that is,

$$3 + 3 + 3 + 3 + 3 + 3.$$

The sum 18 is called the *product* of 6 times 3. This method of associating 18 with the pair 6 and 3 is the binary operation defined on the set of whole numbers called *multiplication*.

Figure 1.   *Product* 6 × 3.

In general, if *a* is any natural number greater than 1 and if *b* is any whole number, then we define the product of *a* times *b*, symbolized by *a* × *b*, *ab*, or *a*·*b*, to be the *sum* of *a* numbers each of which is *b*. Thus,

$$a \times b = \underbrace{b + b + b + \cdots + b + b.}_{a \text{ terms}}$$

Since the set of whole numbers is closed with respect to addition, the set of whole numbers is also closed with respect to multiplication.

*Examples*

1. $4 \times 2 = 2 + 2 + 2 + 2 = 8$.
2. $5 \times 3 = 3 + 3 + 3 + 3 + 3 = 15$.
3. $6 \times 0 = 0 + 0 + 0 + 0 + 0 + 0 = 0$.
4. $4 \times 1 = 1 + 1 + 1 + 1 = 4$.

The first thing we should notice is that $6 \times 3 = 3 \times 6$. This can be shown by observing that

$$3 + 3 + 3 + 3 + 3 + 3 = 6 + 6 + 6.$$

Another way to see that $6 \times 3 = 3 \times 6$ would be to look at Figure 2 below.

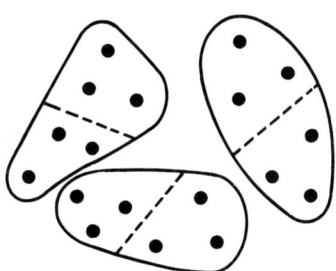

Figure 2. *Product* $3 \times 6$.

Still another way to show that $6 \times 3 = 3 \times 6$ is to consider the following rectangular arrangement of dots.

```
    *  *  *  *  *  *
    *  *  *  *  *  *
    *  *  *  *  *  *
```

We can see that three rows with six dots each gives the same total as six columns with three dots each. One important feature of this technique to show that $6 \times 3 = 3 \times 6$ is that it generalizes. It can be

## 4.1 Multiplication

used to show that $a \times b = b \times a$ for whole numbers $a$ and $b$ where $a > 1$ and $b > 1$.

From the definition of multiplication, $0 \times 6$ has not been defined. (We cannot talk about adding 0 sixes since addition is a binary operation.) Since $6 \times 0 = 0$, in order to preserve closure and the commutative property of multiplication we *define* $0 \times 6$ to be zero. In general, for any whole number $a$, we define $0 \times a$ to be zero; thus,

$$a \times 0 = 0 \times a = 0.$$

Similarly, since addition is a binary operation, we cannot talk about the sum, say, of one 7. However, since $7 \times 1 = 7$, we *define* $1 \times 7$ to be 7. In general, for any whole number $a$, we define $1 \times a$ to be $a$; thus,

$$a \times 1 = 1 \times a = a.$$

**Property 1:** Closure property of multiplication. For any two whole numbers $a$ and $b$, the product $a \times b$ is a unique whole number.

**Property 2:** Commutative property of multiplication. For any pair of whole numbers $a$ and $b$, $a \times b = b \times a$.

**Property 3:** Multiplicative identity. Let $a$ be any whole number. Then $a \times 1 = 1 \times a = a$ and 1 is called the *multiplicative identity* for the set of whole numbers.

If we recall the basic properties of addition, we see that the only similar property not listed above is the associative property. We first ask ourselves, is $3 \times (4 \times 6) = (3 \times 4) \times 6$? Since

$$3 \times (4 \times 6) = 3 \times 24 = 72 \quad \text{and} \quad (3 \times 4) \times 6 = 12 \times 6 = 72,$$

we see that the answer to this question is "yes." Of course, one example does not prove that multiplication is an associative operation. Without actually finding the indicated products in our example, we could have shown that the equality held by considering the two configurations in Figure 3. We see in Figure 3 on page 52 that there is the same number of dots in both configurations. On the left, each plane (group) consists of $4 \times 6$ dots. Since there are three planes, the total number of dots is

$$3 \times (4 \times 6).$$

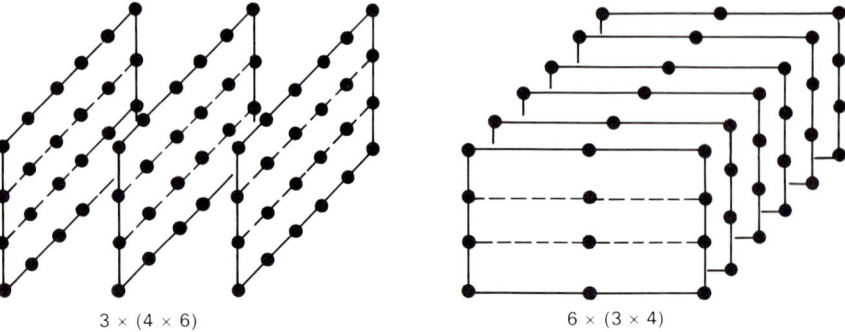

$3 \times (4 \times 6)$          $6 \times (3 \times 4)$

Figure 3. *Associative property*.

On the right, the front plane contains $3 \times 4$ dots. Since there are 6 such planes, the total number of dots is $6 \times (3 \times 4)$, or

$$(3 \times 4) \times 6.$$

The procedure generalizes to show that if $a$, $b$, and $c$ are whole numbers then $a \times (b \times c) = (a \times b) \times c$.

**Property 4:** Associative property of multiplication. For any whole numbers $a$, $b$, and $c$, $a \times (b \times c) = (a \times b) \times c$.

Let us discuss one more property of addition and multiplication which is important enough to deserve a distinguishing name. Consider the product $4 \times (5 + 7)$. From our definition of multiplication.

$$4 \times (5 + 7) = (5 + 7) + (5 + 7) + (5 + 7) + (5 + 7).$$

Using the associative and commutative properties of addition, we obtain

$$4 \times (5 + 7) = (5 + 5 + 5 + 5) + (7 + 7 + 7 + 7).$$

Again, by definition of multiplication,

$$4 \times (5 + 7) = (4 \times 5) + (4 \times 7).$$

In general, if $a$, $b$, and $c$ are whole numbers, we can prove that $a(b + c) = (ab) + (ac)$.

### DISTRIBUTIVE PROPERTY

```
        4(5 + 7)              =     (4 × 5)        +        (4 × 7)
  * * * * * * * * * * * *          * * * * *             * * * * * * *
  * * * * * * * * * * * *          * * * * *             * * * * * * *
  * * * * * * * * * * * *    =     * * * * *       +     * * * * * * *
  * * * * * * * * * * * *          * * * * *             * * * * * * *
```

## 4.1 Multiplication

Notice that $(ab) + (ac)$ represents the sum of the products $ab$ and $ac$. To avoid an unnecessary use of parentheses, mathematicians agree that multiplication takes precedence over addition. (This agreement is for convenience and has nothing to do with the relative importance of the two operations.) Therefore, we usually write $ab + ac$ instead of $(ab) + (ac)$.

**Property 5:** Distributive property. If $a$, $b$, and $c$ are whole numbers, then $a(b + c) = ab + ac$.

Let $x$ be any *natural* number. Any number which is the product of 3 and $x$ is called a *multiple* of 3. For example, $3 \times 1 = 3$, $3 \times 2 = 6$, $3 \times 3 = 9$, and $3 \times 4 = 12$ are (the first four) multiples of 3. In general, for a given *natural number* $a$ the number $xa$ where $x$ is a natural number is called a multiple of $a$. Notice that in the set of multiples of 3 there is not a greatest element; however, there is a least element, and it is 3. We should specifically point out that zero is *not* the least multiple of 3 since by definition it is not a multiple of 3. In fact, by our definition, zero is not a multiple of any natural number.

### Exercises I

1. Find each of the indicated products. (a) $3 \times 8$. (b) $7 \times 8$. (c) $9 \times 6$. (d) $4 \times 0$. (e) $7 \times 2$.
2. Make a multiplication table for 0, 1, 2, 3, 4, 5, 6, 7, 8, and 9 similar to the addition table on page 25.
3. For the multiplication table constructed in Exercise 2, answer each of the following questions.
   (a) What element is in the first row and first column of the entries?
   (b) What can you say about the second row compared to the headline?
   (c) Are the entries symmetrically placed with respect to the main diagonal?
4. In each case, determine what number $N$ represents.
   (a) $9 \times 6 = N$.  (b) $6 \times 0 = N$.  (c) $8 \times 7 = N$.
   (d) $6 \times N = 48$.  (e) $7 \times N = 0$.  (f) $9 \times N = 72$.
   (g) $7 \times N = 28$.  (h) $5 \times N = 45$.  (i) $6 \times 7 = N$.
5. (a) Is there a unique whole number $N$ such that $7 \times N = 21$?

(b) Is there a unique whole number $N$ such that $7 \times N = 32$?

(c) Is there a unique whole number $N$ such that $0 \times N = 0$?

6. (a) List ten multiples of 4.   (b) List six multiples of 5.
7. (a) Does the set of multiples of 4 have a greatest element?
   (b) Does the set of multiples of 5 have a least element?
   (c) If $S$ is the set of multiples of 4 and $T$ is the set of multiples of 5, list three elements in $S \cap T$.
8. Let $A$ be the set of multiples of 3 and let $B$ be the set of multiples of 4.
   (a) List ten elements in the set $A \cap B$.
   (b) What is the least element in $A \cap B$?
9. (a) List the first ten multiples of 6.
   (b) List the first ten multiples of 15.
   (c) If $A$ is the set of multiples of 6 and $B$ is the set of multiples of 15, list three elements in $A \cap B$.
   (d) What is the least element in $A \cap B$?
10. Prove for whole numbers $a$, $b$, and $c$ that $(a + b)c = ac + bc$.

## Exercises II

1. Let $a$ and $b$ be natural numbers.
   (a) If $A$ is the set of multiples of $a$ and $B$ is the set of multiples of $b$, how do we know that $A \cap B$ is not the empty set?
   (b) What property tells us that $A \cap B$ has a least element?
2. Let $a$, $b$, $c$, and $d$ be whole numbers. Use the basic properties of addition and multiplication to prove that $a(b + c + d) = ab + ac + ad$.
3. Prove that if $u$ and $v$ are each multiples of a natural number $a$, then $u + v$ is a multiple of $a$.
4. (a) Let $S$ be the set of all multiples of 4 which are less than or equal to 32. List the elements in $S$.
   (b) What is the greatest element in $S$?
   (c) If $t$ is the greatest element in $S$, what is $32 - t$?
   (d) What is $n(S)$?
5. (a) Let $S$ be the set of all the multiples of 6 which are less than or equal to 47. List the elements in $S$.

(b) What is the greatest element in $S$?
(c) If $t$ is the greatest element in $S$, what is $47 - t$?
(d) What is $n(S)$?
6. Let $A$ be a set with $a$ elements and let $b$ be a natural number less than $a$. Interpret each of the following statements in terms of sets and the grouping of elements in sets.
   (a) There are $n$ multiples of $b$ less than or equal to $a$.
   (b) If $qb$ is the greatest multiple of $b$ less than or equal to $a$, then $a - qb < b$.
7. Let $a$ and $b$ be whole numbers where $a < b$. Is $ac < bc$ for every whole number $c$?
8. Let $a$ and $b$ be whole numbers where $a < b$. Prove if $c$ is any *natural number*, then $ac < bc$.
9. Let $a$, $b$, and $c$ be natural numbers. Prove if $a$ is a multiple of $c$ then $ab$ is a multiple of $c$.
10. Let $a$, $b$, and $c$ be natural numbers. Is it true that if $ab$ is a multiple of $c$ then either $a$ or $b$ is a multiple of $c$?

## 4.2 Techniques for Multiplication

One method to find the product $27 \times 38$ is to find the sum of twenty-seven numbers each of which is 38. An even less efficient method would be to find the sum of thirty-eight numbers each of which is 27. Obviously, we prefer a shorter technique; one can readily be found by considering a few facts we have already learned.
Since $27 = 20 + 7$,

$$(27)(38) = (20 + 7)38$$
$$= (20)(38) + (7)(38).$$

Now, let us first consider finding the product $7 \times 38$.

$$\begin{aligned}7 \times 38 &= 7(30 + 8) \\ &= (7)(30) + (7)(8) \\ &= [7 \times (3 \times 10)] + 56 \\ &= [(7 \times 3) \times 10] + (50 + 6) \\ &= (21 \times 10) + (5 \times 10) + 6 \\ &= (26 \times 10) + 6 \\ &= 266.\end{aligned}$$

Notice that $7 \times 30 = 7 \times (3 \times 10) = (7 \times 3) \times 10$, or 21 tens.

Observing this, we can shorten the technique as follows. Multiply 7 times 8 and write down this product 56. Then, multiply 7 times 30 and write the product 210 below 56 in a column, and, finally, find the sum of 56 and 210. The procedure is exhibited as follows.

$$\begin{array}{r} 38 \\ 7 \\ \hline 56 \\ 210 \\ \hline 266 \end{array}$$

The technique is shortened still further by observing for the product $7 \times 8 = 56$ that the 5 tens is added to the $7 \times 3 = 21$ tens. In other words, we can "carry" the 5 (tens) to be added to the product $7 \times 30$ and save writing another line. This is exhibited as follows.

$$\begin{array}{r} 5 \\ 38 \\ 7 \\ \hline 266 \end{array}$$

Now, consider the product $20 \times 38$. This is $(2 \times 10) \times 38 = 10 \times (2 \times 38)$. Multiply 2 times 38 in the manner just discussed.

$$\begin{array}{r} 1 \\ 38 \\ 2 \\ \hline 76 \end{array}$$

We find that $20 \times 38 = 10 \times (2 \times 38) = 10 \times 76$, or 760. Thus,

$$\begin{array}{r} 38 \\ 27 \\ \hline 266 \\ 760 \\ \hline 1026 \end{array}$$

This procedure can be generalized for numbers with any number of digits. As an example, consider the following.

Multiply:  384
          263

*4.2 Techniques for Multiplication*      57

| **Step 1.** | **Step 2.** | **Step 3.** |
|---|---|---|
| 2 1 | 5 2 | 1 |
| 3 8 4 | 3 8 4 | 3 8 4 |
|     3 |     6 |     2 |
| 1 1 5 2   units | 2 3 0 4   tens | 7 6 8   hundreds |

**Step 4.** The product is the sum of the products found in Steps 1, 2, and 3. Thus, the product is

$$\begin{array}{r} 1\,1\,5\,2 \\ 2\,3\,0\,4\,0 \\ 7\,6\,8\,0\,0 \\ \hline 1\,0\,0{,}9\,9\,2 \end{array}$$

Of course, the next step is to combine Steps 1, 2, 3, and 4 and perform the multiplication as follows.

$$\begin{array}{r} 3\,8\,4 \\ 2\,6\,3 \\ \hline 1\,1\,5\,2 \\ 2\,3\,0\,4\,0 \\ 7\,6\,8\,0\,0 \\ \hline 1\,0\,0{,}9\,9\,2 \end{array}$$

In a final effort to shorten our work, we often omit writing the "unnecessary" zeros and set the intermediate products over one position to the left each time. Therefore, the final procedure would be as follows.

$$\begin{array}{r} 3\,8\,4 \\ 2\,6\,3 \\ \hline 1\,1\,5\,2 \\ 2\,3\,0\,4 \\ 7\,6\,8 \\ \hline 1\,0\,0{,}9\,9\,2 \end{array}$$

Since multiplication is a commutative operation, one method to check this product is to use our *algorithm* (technique) to find the product 384 × 263. Here, it is necessary to distinguish between the products 263 × 384 and 384 × 263. In order to make this distinction, we give names to the numbers in such a product. In the product 384 × 263, the number 384 is called the multiplier and the number 263 is called the multiplicand. Let us check our previous product by performing the multiplication in the other order.

Multiply:        2 6 3   (multiplicand)
                 3 8 4   (multiplier)
                ───────
                 1 0 5 2
                 2 1 0 4
                 7 8 9
                ─────────
                 1 0 0,9 9 2   (product)

It should not be unexpected that we also have a casting-out-nines check for multiplication. It is exhibited as follows.

Multiply:    3 8 4  —+→ 15  —+→  6
             2 6 3  —+→ 11  —+→  2
            ───────                    —+→
             1 1 5 2           Product 1 2  —+→ 3
             2 3 0 4
             7 6 8
            ───────
             1 0 0,9 9 2  —+→ 21  —+→ 3

There are probably only two more examples which we need to consider for the multiplication algorithm; both involve zeros in the multiplier. If zero is in the multiplier, we can either multiply by 0 and proceed as usual, or we can omit the zeros through proper positioning of the other digits.

Consider the following examples.

## *Examples*

**Method 1.**  (a)    6 8 2          (b)    2 8 4
                     1 0 3                  2 3 0
                    ───────                ───────
                    2 0 4 6                0 0 0
                    0 0 0                  7 5 2
                    6 8 2                  5 6 8
                    ─────────              ─────────
                    7 0,2 4 6              6 4,3 2 0

**Method 2.**  (a)    6 8 2          (b)    2 8 4
                     1 0 3                  2 3 0
                    ───────                ───────
                    2 0 4 6                7 5 2
                    6 8 2                  5 6 8
                    ─────────              ─────────
                    7 0,2 4 6              6 4,3 2 0

The algorithm for multiplication discussed in this section is generally considered the standard technique for multiplying natural numbers. Of course, there are several slight variations which are often employed. In Chapter 5, we shall discuss a completely different algorithm for finding such products.

## 4.3 Factors

### Exercises I

Find the products of each pair of numbers in Exercises 1 through 12 and check your answers by the casting-out-nines technique.

1. 5 2 6
      1 8

2. 6 1 8
      1 9

3. 4 9 2
      1 6

4. 6 5 1
      8 3

5. 8 2 2
      3 5

6. 4 8 7
      6 4

7. 3 2 7
      2 0 6

8. 2 8 9
      3 7 0

9. 8 2 5
      6 0 0

10. 6,2 6 7
       2,4 2 1

11. 8,0 2 6
       3,7 0 8

12. 4,2 7 1
       6,0 0 8

### Exercises II

Find the products in Exercises 1 through 12 and check your answers by the casting-out-nines technique.

1. 2 8 4
      8 7 1

2. 4 8 6
      4 0 8

3. 1 8 4
      8 3

4. 4 0 0
      8 2 1

5. 9 4 2
      5 0 0

6. 5 0 3
      1 0 1

7. 3,5 7 2
      8 7 3

8. 4,5 7 2
      3 7 1

9. 2 3,6 8 1
      1 0,8 4 6

10. 2 4,5 8 2
       2,0 0 3

11. 3 6 7,8 9 2
       1 0,2 1 2

12. 7 3 2,8 1 3
       4 2 1,6 7 5

## 4.3 Factors

We recall from our definition in Section 4.1 that 6 is a multiple of 3. To express the fact that there is a natural number $x$ such that $3x = 6$, we also say that 3 is a *factor* of 6. In general, for any natural number $b$, a natural number $a$ is called a factor of $b$ if and only if there is a natural number $x$ such that $ax = b$.

For example, 4 is a factor of 28 since there is a natural number $x$, namely 7, such that $4x = 28$.

To express the fact that 4 is a factor of 28 we write $4|28$ (read "4 is a factor of 28", or "4 divides 28"). Since 3 is not a factor of 28, we often write $3 \!\not|\, 28$ (read "3 is not a factor of 28") to express this fact.

For any natural number $a$, since $a \times 1 = 1 \times a = a$, it follows that 1 and $a$ are factors of $a$. If $a \neq 1$, then $a$ must have at least two different factors, namely, itself and 1. Obviously, the only factor of 1 is 1.

### Examples

1. Factors of 6: 1, 2, 3, 6.
2. Factors of 7: 1 and 7.
3. Factors of 8: 1, 2, 4, 8.
4. $2|6$, $3|6$, $3\!\not|\,7$, $6|6$, $4\!\not|\,6$, $4|8$.

In terms of sets, to say that 28 has 4 as a factor is equivalent to stating that a set with 28 objects can be partitioned into disjoint subsets such that each subset contains exactly 4 elements and the union of the disjoint subsets is the original set. In other words, the set can be split up into subsets with 4 elements in each and with no elements left over. Since $4 \times 7 = 28$, we conclude that there are 7 subsets of a set containing 28 elements such that each subset contains exactly 4 elements.

Let $B$ and $C$ be two disjoint sets with $b$ and $c$ the number of elements in each, respectively. If all of the $b$ elements in $B$ can be split into subsets so that each contains $a$ elements, then $a$ is a factor of $b$. Similarly, if all of the $c$ elements in $C$ can be split into subsets so that each contains $a$ elements, then $a$ is a factor of $c$. It should be intuitively clear that it would be possible to split the set $B \cup C$ into subsets so that each contains exactly $a$ elements. Since $n(B \cup C) = b + c$, this means that $a$ is a factor of $b + c$. Let us prove this fact using the definition of factor and the properties of multiplication and addition.

Assume $a|b$ and $a|c$. Then by definition of factor, there are natural numbers $x$ and $y$ such that $ax = b$ and $ay = c$. Notice that $x$ and $y$ need not be different numbers but since they might be we must use different letters to represent them. Now,

## 4.3 Factors

$$b + c = ax + ay$$

and

$$b + c = a(x + y)$$

by the distributive property. Since the sum of two natural numbers is a natural number, $(x + y)$ is a natural number. Therefore, by definition of factor, $a|(b + c)$.

At this point, we have very little machinery to find the factors of a given natural number. In the next chapter, we shall discuss the operation of division and give some tests for finding factors called *divisibility tests*. This will make our task of determining the factors of natural numbers much easier.

### Exercises I

1. List all the factors of 32.
2. List all the factors of 28.
3. List all the factors of 23.
4. List all the factors of 40.
5. List all the factors of 100.

Let $s(n)$ (read "$s$ of $n$") be the sum of *all* the factors of a natural number $n$ and let $d(n)$ be the number of such factors. For example, since all the factors of 6 are 1, 2, 3, and 6, it follows that $s(6) = 1 + 2 + 3 + 6 = 12$ and $d(6) = 4$.

6. (a) What is $s(32)$? (b) What is $d(32)$?
7. (a) What is $s(28)$? (b) What is $d(28)$?
8. (a) What is $s(23)$? (b) What is $d(23)$?
9. (a) What is $s(40)$? (b) What is $d(40)$?
10. (a) What is $s(100)$? (b) What is $d(100)$?

### Exercises II

1. (a) What is $s(5)$? (b) What is $s(6)$? (c) What is $s(30)$?
2. (a) What is $d(5)$? (b) What is $d(6)$? (c) What is $d(30)$?
3. (a) What is $s(4)$? (b) What is $s(15)$? (c) What is $s(60)$?
4. (a) What is $d(4)$? (b) What is $d(15)$? (c) What is $d(60)$?
5. (a) What is $s(7)$? (b) What is $s(11)$? (c) What is $s(77)$?

6. (a) What is $d(7)$? (b) What is $d(11)$? (c) What is $d(77)$?
7. (a) What is $s(8)$? (b) What is $s(10)$? (c) What is $s(80)$?
8. (a) What is $d(8)$? (b) What is $d(10)$? (c) What is $d(80)$?
9. (a) Let $a|b$ and $a|c$. Prove that $a|bc$.
   (b) If $a|bc$, is it true that $a|b$ or $a|c$?
10. Prove if $a|b$ and $b|a$ then $a = b$.

## 4.4 Supplementary Exercises

In Exercises 1 through 7, determine the number $N$ represents.
1. $(4 \times 3) \times 6 = 3 \times (N \times 4)$.
2. $(8 \times N) + 4 = 9 \times 4$.
3. $(6 \times N) + 8 = 83 - 27$.
4. $(7 \times N) - 3 = 4 \times 15$.
5. $(7 + 8) + 21 = 6 + (5 \times N)$.
6. $(18 - 5) + 3 = 4(N + 1)$.
7. $8(N + 2) - (9 + 1) + 22 = (23 + 81) - 44$.
8. (a) List ten multiples of 24. (b) List ten multiples of 20.
9. (a) List all the factors of 24. (b) List all the factors of 20.
10. (a) Let $A$ be the set of multiples of 24 and let $B$ be the set of multiples of 20. List three elements in $A \cap B$.
    (b) Does $A \cap B$ have a greatest element?
11. (a) Let $C$ be the set of factors of 24 and let $D$ be the set of factors of 20. List all the elements in $C \cap D$.
    (b) What is the least element in $C \cap D$?
    (c) What is the greatest element in $C \cap D$?
12. (a) Let $S$ be the set of multiples of 8 which are less than or equal to 93. List the elements in $S$.
    (b) What is the greatest element in $S$?
    (c) If $t$ is the greatest element in $S$, what is $93 - t$?
    (d) What is $n(S)$?
    (e) In terms of sets, explain what information is given by the answers to parts (c) and (d).
13. (a) Let $S$ be the set of multiples of 7 which are less than or equal to 84. List the elements in $S$.
    (b) What is the greatest element in $S$?
    (c) If $t$ is the greatest element in $S$, what is $84 - t$?
    (d) What is $n(S)$?
    (e) In terms of sets, explain what information is given by the answers to parts (c) and (d).

## 4.4 Supplementary Exercises

14. (a) What is $s(14)$? (b) What is $s(15)$? (c) What is $s(210)$?
    (d) What is $d(14)$? (e) What is $d(15)$? (f) What is $d(210)$?
15. (a) What is $s(3)$? (b) What is $s(11)$? (c) What is $s(33)$?
    (d) What is $d(3)$? (e) What is $d(11)$? (f) What is $d(33)$?
16. (a) What is $s(20)$? (b) What is $s(7)$? (c) What is $s(140)$?
    (d) What is $d(20)$? (e) What is $d(7)$? (f) What is $d(140)$?

In Exercises 17 through 20, find the sum, difference, and product of each pair of numbers and check your answers by the casting-out-nines technique.

17. 23,846 and 12,468.
18. 308,261 and 456,221.
19. 432,966 and 400,325.
20. 821,344 and 451,083.

ac = b  ● b ÷ a = c
───────────────────
42 ÷ 6 = 7  ●  x⟌xx

# DIVISION AND INCOMPLETE QUOTIENTS

## 5.1 Division and Divisibility Tests

The four basic operations of arithmetic are addition, subtraction, multiplication, and division; these are called the rational operations. Division is defined in terms of multiplication in much the same way as subtraction is defined in terms of addition. The student should compare these two definitions.

Let $a$ and $b$ be any numbers. *If there exists a unique number $c$ such that $ac = b$, then $c$ is called the* quotient of $b$ divided by $a$. Symbolically,

$$b \div a = c \quad \text{if and only if} \quad ac = b.$$

For example, since $6 \times 7 = 42$, it follows that $42 \div 6 = 7$. Of course, since $7 \times 6 = 42$, it also follows that $42 \div 7 = 6$. For the quotient $b \div a$ where $a$ and $b$ are whole numbers, the number $b$ is called the dividend and $a$ is called the divisor.

We encounter some of the same difficulties with division for whole numbers that we did with subtraction. First, the set of whole numbers is not closed with respect to division. Since, for example, there is no whole number $x$ such that $3x = 7$, the quotient $7 \div 3$ does not exist in the set of whole numbers. Also, it is immediately apparent that division is not a commutative operation; observe that $6 \div 3 \neq 3 \div 6$.

Consider $0 \div a$ where $a$ is not zero. Since there is a unique number $x$, namely zero, such that $ax = 0$ it follows from the definition of division that

$$\text{if } a \neq 0 \quad \text{then} \quad 0 \div a = 0.$$

Since *any* number times zero is zero (for example, $0 \times 6 = 0$ and $0 \times 8 = 0$), there is no *unique* number $x$ such that $0(x) = 0$; thus, $0 \div 0$ is not defined. Also, since there is *no* number $x$ such that, say, $0(x) = 6$, the quotient $6 \div 0$ is not defined. In summary, division by zero is not defined.

For the set of natural numbers, we should observe how the ideas of factor and multiple are related to the operation of division. Since $2 \times 3 = 6$, 6 is a multiple of 2, 2 is a factor of 6, and the quotient $6 \div 2$ is another factor of 6.

A natural number is called even if and only if it is a multiple of 2; it is called odd if and only if it is not a multiple of 2. The first fifteen multiples of 2 are

$$2, 4, 6, 8, 10, 12, 14, 16, 18, 20, 22, 24, 26, 28, 30.$$

We notice that the multiples of 2 which are greater than 8 have 0, 2, 4, 6, or 8 as their last digit in the positional representation of the number. Not only is it true that any even natural number has a last digit of 0, 2, 4, 6, or 8, but the converse is also true; that is, every natural number with a last digit of 0, 2, 4, 6, or 8 is a multiple of 2. Of course, to say that a natural number is a multiple of 2 is to say it has 2 as a factor. The statement:

> A natural number has 2 as a factor if and only if the last digit in the positional representation for the number is 0, 2, 4, 6, or 8.

is called a *divisibility test* for 2.

One way to find if 5 is a factor of 325 is to find the natural number $x$ such that $5x = 325$. Actually, most of us know a much easier way to find out if 5 is a factor of 325; it is to use the divisibility test for 5. The divisibility test for 5 is:

> A natural number has 5 as a factor if and only if the last digit in the positional representation for the number is 0 or 5.

For example, both 325 and 480 have 5 as a factor, but 743 does not have 5 as a factor.

Another familiar divisibility test is:

> A natural number has 10 as a factor if and only if the last digit in the positional representation for the number is 0.

Although we do not prove the validity of the divisibility tests until Chapter 12, let us state some of the other tests and see how they are used.

## 5.1 Division and Divisibility Tests

Consider the number 378,425,664. From the three divisibility tests already discussed, we know that this number has 2 as a factor but does not have 5 or 10 as factors. Let us determine which of the numbers 3, 4, 5, 6, 8, 9, 11, 12, 15, and 18 are factors of 378,425,664. First, we find the sum of all the digits in the number 378,425,664.

$$
\begin{array}{rr}
\text{Add:} & 3 \\
& 7 \\
& 8 \\
& 4 \\
& 2 \\
& 5 \\
& 6 \\
& 6 \\
& 4 \\ \hline
\textbf{Sum} & \textbf{45}
\end{array}
$$

Using both the digits and sum of the digits in the given number, we can deduce a number of important facts. They are as follows.

1. Since the resulting sum, 45, has 9 as a factor, the original number is divisible by 9.
2. Since the resulting sum has 3 as a factor, the original number is divisible by 3.
3. Since the last *two digits*, 64, of the given number represents a number divisible by 4, the given number is divisible by 4.
4. Since the last *three digits*, 664, of the given number represents a number divisible by 8, the given number is divisible by 8.
5. Since the given number is divisible by 2 and 9, it is divisible by 18.
6. Since the given number is divisible by 2 and 3, it is divisible by 6.
7. Since the given number is divisible by 3 and 4, it is divisible by 12.
8. Consider the digits in the odd-numbered positions from the right in the given number: 37**8**,4**2**5,**6**64. Find their sum.

$$
\begin{array}{rr}
\text{Add:} & 3 \\
& 8 \\
& 2 \\
& 6 \\
& 4 \\ \hline
\textbf{Sum} & \textbf{23}
\end{array}
$$

Consider the digits in the even-numbered positions from the right: 3**7**8,**4**25,6**6**4. Find their sum.

$$\begin{array}{rr} \text{Add:} & 7 \\ & 4 \\ & 5 \\ & 6 \\ \hline \text{Sum} & 22 \end{array}$$

Since the difference of the larger and smaller of these two sums $(23 - 22 = 1)$ is neither zero nor has 11 as a factor, the given number does not have 11 as a factor. (Here, as with the other tests, the converse is true; that is, if the difference were zero or a multiple of 11, the number would have 11 as a factor.)
9. Since the given number does not have 5 as a factor, it cannot have any multiple of 5, such as 15 or 40, as a factor.

In Chapter 12, we show by rather elementary methods that these tests to check the divisibility of 378,425,664 by the natural numbers 2, 3, 4, 5, 6, 8, 9, 10, 11, 12, 15, and 18 are valid for any given natural numbers. Let us formally state these divisibility tests.

## Divisibility Tests

1. A natural number has 2 as a factor if and only if its last digit is 2, 4, 6, 8, or 0.
2. A natural number has 3 as a factor if and only if the sum of its digits has 3 as a factor.
3. A natural number has 4 as a factor if and only if the last two digits form a number having 4 as a factor.
4. A natural number has 5 as a factor if and only if the last digit is 0 or 5.
5. A natural number has 6 as a factor if and only if it has 2 and 3 as factors.
6. A natural number has 8 as a factor if and only if the last three digits form a number having 8 as a factor.
7. A natural number has 9 as a factor if and only if the sum of its digits has 9 as a factor.
8. A natural number has 10 as a factor if and only if the last digit is 0.
9. A natural number has 11 as a factor if and only if the difference of the sums of the digits in the even-numbered and odd-numbered positions is zero or a multiple of 11.

## 5.1 Division and Divisibility Tests

10. A natural number has 12 as a factor if and only if it has 3 and 4 as factors.
11. A natural number has 15 as a factor if and only if it has 3 and 5 as factors.
12. A natural number has 18 as a factor if and only if it has 2 and 9 as factors.

### Exercises I

1. Use the multiplication table and the definition of division to find the following quotients, if they exist.

   (a) $72 \div 8$.  (b) $36 \div 9$.  (c) $0 \div 7$.  (d) $0 \div 0$.
   (e) $83 \div 6$.  (f) $54 \div 6$.  (g) $7 \div 0$.  (h) $64 \div 8$.

2. (a) Discuss what is wrong with the following "divisibility test." A natural number is divisible by 40 if and only if it is divisible by 4 and 10.
   (b) State a proper divisibility test for 40.
3. Use divisibility tests to find which of the numbers 2, 3, 4, 5, 6, 8, 9, 10, 11, 12, 15, 18, and 40 are factors of 8,440.
4. Use divisibility tests to find which of the numbers 2, 3, 4, 5, 6, 8, 9, 10, 11, 12, 15, 18 and 40 are factors of 87,120.
5. Use divisibility tests to find which of the numbers 2, 3, 4, 5, 6, 8, 9, 10, 11, 12, 15, 18, and 40 are factors of 4,759.
6. Use divisibility tests to find which of the numbers 2, 3, 4, 5, 6, 8, 9, 10, 11, 12, 15, 18, and 40 are factors of 61,843,320.
7. Use divisibility tests to find which of the numbers 2, 3, 4, 5, 6, 8, 9, 10, 11, 12, 15, 18, and 40 are factors of 737,880.
8. Use divisibility tests to find which of the numbers 2, 3, 4, 5, 6, 8, 9, 10, 11, 12, 15, 18, and 40 are factors of 4,820,655,852.
9. Use divisibility tests to find which of the numbers 2, 3, 4, 5, 6, 8, 9, 10, 11, 12, 15, 18, and 40 are factors of 4,005,830,664.
10. Use divisibility tests to find which of the numbers 2, 3, 4, 5, 6, 8, 9, 10, 11, 12, 15, 18, and 40 are factors of 163,179,720.

## Exercises II

1. Let $a,bcd$ be a four-digit number. Answer each question in the following.
$$\begin{aligned}a,bcd &= 1000a + 100b + 10c + d &&\text{(1) Why?}\\ &= (999 + 1)a + (99 + 1)b + (9 + 1)c + d &&\text{(2) Why?}\\ &= 999a + a + 99b + b + 9c + c + d &&\text{(3) Why?}\\ &= (999a + 99b + 9c) + (a + b + c + d) &&\text{(4) Why?}\\ &= 9(111a + 11b + c) + (a + b + c + d) &&\text{(5) Why?}\end{aligned}$$
If the sum of the digits $a + b + c + d$ has 9 as a factor, then the number $a,bcd$ has 9 as a factor. (6) Why?

2. Using the divisibility tests, list all the factors of 6,060.
3. (a) List all the factors of 20.
   (b) List all the factors of 45.
   (c) List ten multiples of 20.
   (d) List ten multiples of 45.
4. (a) What is the greatest factor that 20 and 45 have in common?
   (b) What is the least multiple that 20 and 45 have in common?
5. If $G$ is the greatest factor that 20 and 45 have in common and if $L$ is the least multiple that 20 and 45 have in common, what is the product $G \times L$? (See Exercise 4.)
6. For $G$ and $L$ in Exercise 5, is it true that $G \times L = 20 \times 45$?
7. (a) List all the factors of 18.
   (b) List all the factors of 30.
   (c) List ten multiples of 18.
   (d) List ten multiples of 30.
8. (a) What is the greatest factor that 18 and 30 have in common?
   (b) What is the least multiple that 18 and 30 have in common?
9. If $G$ is the greatest factor that 18 and 30 have in common and if $L$ is the least multiple that 18 and 30 have in common, what is the product $G \times L$? (See Exercise 8.)
10. (a) For $G$ and $L$ of Exercise 9, is it true that $G \times L = 18 \times 30$?
    (b) Is it true that $(18 \times 30) \div G = L$?

## 5.2 The Division Algorithm

From the definition of division, since $4 \times 7 = 28$ we conclude that $28 \div 4 = 7$. In terms of sets, this means that a set containing 28 elements can be split up into 7 subsets each of which contains exactly 4 elements. As we can interpret multiplication of whole numbers as repeated addition, we can interpret division of whole numbers as repeated subtraction. Consider the following.

| Number of Subtractions | 1 | 2 | 3 | 4 | 5 | 6 | 7 |
|---|---|---|---|---|---|---|---|
| | 28 | 24 | 20 | 16 | 12 | 8 | 4 |
| | 4 | 4 | 4 | 4 | 4 | 4 | 4 |
| | 24 | 20 | 16 | 12 | 8 | 4 | 0 |

The quotient 7 is the number of repeated subtractions necessary to get zero as a difference.

As we learned in the last section, the set of whole numbers is not closed with respect to division. For example, there is no whole number $x$ such that $35 \div 4 = x$. In terms of sets, this means we cannot separate a set containing 35 elements into subsets such that each subset will have exactly 4 elements. In terms of our interpretation of division as repeated subtraction, this means we do not ever obtain zero as a difference in the associated sequence of repeated subtractions.

For a set with 35 elements, the greatest number of subsets which have exactly 4 elements is 8 and the number of elements left over is 3. We notice that 8 subsets each containing 4 elements account for 32 elements; also, 32 is the greatest multiple of 4 less than 35.

Let $a$ and $b$ be natural numbers where $a \geq b$. Since $b$ is a multiple of $b$, the set of multiples of $b$ which are less than or equal to $a$ is not empty. If $qb$ is the greatest multiple of $b$ less than or equal to $a$, then the number $q$ is called the incomplete quotient and $a - qb$ is called the remainder in the *incomplete division* of $a$ by $b$. If $a - qb = 0$, then $a = qb$ and $a \div b = q$. Hence, if the remainder is zero, the incomplete quotient is the quotient $a \div b$. Our immediate task is to present an algorithm which will determine the two numbers $q$ and $r$ for any given natural numbers $a$ and $b$ where $a \geq b$.

Given $a = 678$ and $b = 23$. To find the numbers $q$ and $r$ such that $23q$ is the greatest multiple of 23 less than or equal to 678 and where $r = 678 - 23q$, we could proceed by listing the successive multiples of 23 until we found one which exceeded 678.

$$1 \times 23 = 23$$
$$2 \times 23 = 46$$
$$3 \times 23 = 69$$
$$4 \times 23 = 92$$
$$5 \times 23 = 115$$
$$6 \times 23 = 138$$
$$\ldots$$
$$\ldots$$
$$25 \times 23 = 575$$
$$26 \times 23 = 598$$
$$27 \times 23 = 621$$
$$28 \times 23 = 644$$
$$29 \times 23 = 667$$
$$30 \times 23 = 690$$

Thus, $29 \times 23$ is the greatest multiple of 23 less than or equal to 678 and $q = 29$. Since $678 - (29 \times 23) = 678 - 667 = 11$, we see that $r = 11$. Notice that $r$ is a whole number less than the divisor 23. The remainder must be less than 23 or another 23 could have been subtracted.

Performing repeated subtractions and listing successive multiples are generally quite inefficient methods to find $q$ and $r$. A more efficient method would be to proceed as follows.

Our experience with multiplication tells us that

$$20 \times 23 < 678$$

and

$$30 \times 23 > 678.$$

The first inequality shows us that the greatest multiple of 23 less than or equal to 678 is at least as large as $20 \times 23$. The second inequality shows us that the greatest multiple of 23 less than or equal to 678 is less than $30 \times 23$. Since $20 \times 23 = 460$,

$$\begin{array}{rl} \text{Subtract:} & 678 \\ & \underline{460} = 20 \times 23 \\ \text{Difference} & 218 \end{array}$$

Suppose we do not recognize that $9 \times 23$ is less than 218; we might proceed as follows.

## 5.2 The Division Algorithm

$$\begin{array}{rl}
 & 678 \\
 & \underline{460} = 20 \times 23 \\
\text{Difference} & 218 \\
 & \underline{69} = 3 \times 23 \\
\text{Difference} & 149 \\
 & \underline{69} = 3 \times 23 \\
\text{Difference} & 80 \\
 & \underline{69} = 3 \times 23 \\
\text{Difference} & 11 \quad \text{Remainder}
\end{array} \right\} 29 \times 23$$

Thus, $678 - (29)(23) = 11$, $q = 29$, and $r = 11$. Although this technique is not the standard algorithm to find $q$ and $r$, it is quite correct.

Let us now find numbers $q$ and $r$ such that $7846 = 231q + r$ where $r \geq 0$ and $r < 231$. The first method we use is the one just described and the second method is a variation of the first.

**Method I**

$$\begin{array}{l}
7846 \\
\underline{6930} = 30 \times 231 \\
916 \\
\underline{693} = 3 \times 231 \\
223
\end{array}$$

**Method II**

$$\begin{array}{r|l}
231 \overline{)7846} & \\
6930 & 30 \\
\hline
916 & \\
693 & 3 \\
\hline
r = 223 & 33 = q
\end{array}$$

Thus, $q = 33$ and $r = 223$.

Let $a$ and $b$ be natural numbers such that $a \geq b$. We can prove that there exist unique whole numbers $q$ and $r$ such that $a = qb + r$ and $r \leq b$. Although it is a misnomer, this *fact* is generally called the *division algorithm*. The technique to find $q$ and $r$ is often called *long division*. We give several variations of long division to find $q$ and $r$ in the incomplete division of 80,336 by 378; the student should study each carefully.

**Method I.**

$$\begin{array}{r|l}
378 \overline{)80{,}336} & \\
75{,}600 & 200 \\
\hline
4{,}736 & \\
3{,}780 & 10 \\
\hline
956 & \\
756 & 2 \\
\hline
r = 200 & 212 = q
\end{array}$$

**Method II.**

$$\begin{array}{r} 2 \\ 1\,0 \\ 2\,0\,0 \\ 378\,\overline{)80{,}336} \quad q = 212 \\ 75{,}600 \\ \hline 4{,}736 \\ 3{,}780 \\ \hline 956 \\ 756 \\ \hline r = 200 \end{array}$$

**Method III.**

$$\begin{array}{r} 212 = q \\ 378\,\overline{)80{,}336} \\ 76{,}600 \\ \hline 4{,}736 \\ 3{,}780 \\ \hline 956 \\ 756 \\ \hline 200 = r \end{array}$$

**Method IV.**

$$\begin{array}{r} 212 = q \\ 378\,\overline{)80{,}336} \\ 756 \\ \hline 473 \\ 378 \\ \hline 956 \\ 756 \\ \hline 200 = r \end{array}$$

A long division problem can be checked by finding the product of the divisor and the incomplete quotient and then adding the remainder to this product; the sum should be equal to the dividend. In the last example, we find that

$$(378 \times 212) + 200 = 80{,}336;$$

hence, 212 is the incomplete quotient and 200 is the remainder.

### Exercises I

In Exercises 1 through 8, perform the indicated long divisions to

## 5.2 The Division Algorithm

obtain the incomplete quotients and remainders. Check your answers.

1. $36\overline{)5,072}$
2. $41\overline{)8,468}$
3. $72\overline{)3,502}$
4. $268\overline{)4,831}$
5. $323\overline{)46,671}$
6. $428\overline{)869,428}$
7. $671\overline{)891,483}$
8. $841\overline{)1,269,242}$
9. Is there a natural number $x$ such that $478x = 8,216,348$?
10. (a) Use the divisibility tests to show that 18 is a factor of 4,673,520.
    (b) Find the natural number $x$ such that $18x = 4,673,520$.

### Exercises II

1. Let $a$ and $b$ be natural numbers where $a < b$. Prove that there are whole numbers $q$ and $r$ such that $a = qb + r$ where $r \leq b$. (Note: A proof of the existence of $q$ and $r$ where $a \geq b$ is given in Chapter 12.)
2. (a) Use the divisibility tests to show that 11 is a factor of 34,287.
   (b) Find the natural number $x$ such that $11x = 34,287$.
3. (a) Use the divisibility tests to show that 18 is a factor of 29,070.
   (b) Find the natural number $x$ such that $18x = 29,070$.
4. (a) Use the divisibility tests to show that 15 is a factor of 29,070.
   (b) Find the natural number $x$ such that $15x = 29,070$.

In Exercises 5 through 10, perform the indicated long divisions to find the incomplete quotients and remainders. Check your answers.

5. $357\overline{)87,425}$
6. $664\overline{)987,542}$
7. $459\overline{)456,339}$
8. $821\overline{)445,308}$
9. $923\overline{)1,567,920}$
10. $3,200\overline{)8,353,113}$

11. Assume that the quotients $a \div b$ and $c \div d$ are the same. Prove that $ad = bc$.
12. Prove if $(a \div b) \div c = 1$ then $a = bc$.
13. Prove if $a \div (b \div c) = 1$ then $ac = b$.
*14. If each $x$ represents one of the digits 0 through 9 where no

first digit in any number is zero, determine the divisor, dividend, and quotient in the following. (Guessing is not necessary.)

$$\begin{array}{r} x\,x,8\,x\,x \phantom{)} \\ x\,x\,x \overline{\smash{)}x\,x,x\,x\,x,x\,x\,x} \\ \underline{x\,x\,x\phantom{,x\,x\,x,x\,x\,x}} \\ x\,x\,x\,x\phantom{x,x\,x\,x} \\ \underline{x\,x\,x\phantom{x,x\,x\,x}} \\ x\,x\,x\,x\phantom{x} \\ \underline{x\,x\,x\,x\phantom{x}} \\ 0\phantom{x} \end{array}$$

## 5.3 Other Multiplication Algorithms

In Section 4.2, we discussed the standard multiplication algorithm to find a product such as 46 × 128. Let us now discuss a method to find products which requires little more than a knowledge of how to divide by 2, multiply by 2, and add. For example, it does not require a knowledge of such multiplication facts as 6 × 8 = 48. This method is sometimes called the halve-double-sum technique for multiplication.

To find the product 46 × 128, we proceed as follows.

```
     Column A        Column B
       46              128
       23              256  ←┐
       11              512  ←┤
        5             1024  ←┼── Add
        2             2048
        1             4096  ←┘
                     ─────
                     5888   Sum
```

*Explanation of the algorithm.* (1) Write 46 in Column A and write underneath it the quotient 46 ÷ 2, namely 23. (2) Since 23 is an odd number, subtract 1 from 23 to obtain an even number and, then, write the quotient of this number divided by 2. (3) Continue to write in Column A the quotient of the preceding number divided by 2 if the preceding number is even or the quotient of the preceding number less one divided by 2 if the number is odd. (4) Stop writing numbers in Column A when a quotient of 1 is obtained. (5) Now, write 128 at the top of Column B. Then, multiply 128 by 2 and write the product 256 underneath 128. (6) Continue doubling the numbers in Column B until there are as many numbers in this column as there are in Column A. (7) Find the sum of all numbers in Column B which are *opposite the odd numbers in Column A*; this sum is the product 46 × 128.

*Examples*

1. Find the product 39 × 62.

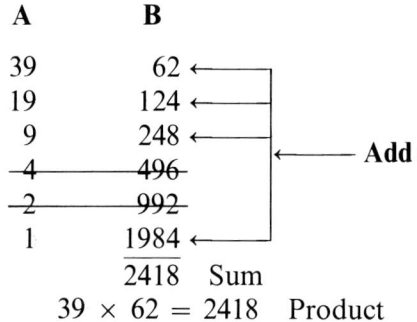

2. Find the product 41 × 52.

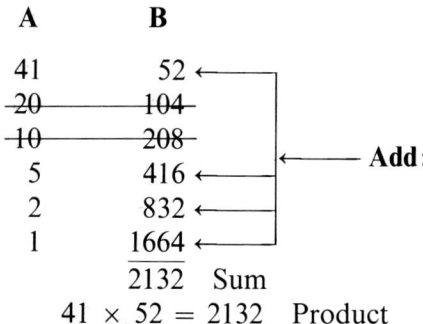

A variation of this procedure is as follows. Divide each number in Column A by 2 and record the incomplete quotient and the remainder. Continue this process until a quotient of 0 is obtained. Then, multiply each *remainder* times the corresponding number in Column B and add these resulting products. This second technique could be done as follows.

|  A | B |
|---|---|
| 46 | |
| 23, $R = 0$ and $0 \times 128 =$ | 0 |
| 11, $R = 1$ and $1 \times 256 =$ | 256 |
| 5, $R = 1$ and $1 \times 512 =$ | 512 |
| 2, $R = 1$ and $1 \times 1024 =$ | 1024 |
| 1, $R = 0$ and $0 = 2048 =$ | 0 |
| 0, $R = 1$ and $1 \times 4096 =$ | 4096 |
| **Sum** | $\overline{5888} = 46 \times 128$ |

Although we postpone a proof of the validity of this technique until Chapter 12, this procedure, though less efficient than the first, should give some insight as to why the halve-double-sum technique works. This second technique is important since there is a completely analogous way to find products where 3 instead of 2 plays the major role. Next, we shall demonstrate the procedure and then it should be obvious that the only advantage of using 2 instead of 3 is that it is easier to multiply by the remainders 0 and 1 than it is to multiply by the remainders 0, 1, and 2 which result when dividing by 3.

Consider two numbers such as 67 and 23. In Column A, write 67 and then underneath write the incomplete quotient and the remainder in the long division of 67 by 3. Below the incomplete quotient 22 write the incomplete quotient and remainder of 22 divided by 3; continue this process until a quotient of zero is obtained. In Column B, write 23 so that it can be multiplied by the first remainder in Column A. Thereafter, write the product of 3 times the preceding number. (See Example 1 below.) Now, multiply each remainder in Column A times the corresponding number in Column B and then find the sum of these products. The sum is the product $67 \times 23$.

*Examples*

1. 

| A | B |
|---|---|
| 67 | |
| 22, $R = 1$ and $1 \times 23 =$ | 23 |
| 7, $R = 1$ and $1 \times 69 =$ | 69 |
| 2, $R = 1$ and $1 \times 207 =$ | 207 |
| 0, $R = 2$ and $2 \times 621 =$ | 1242 |
| | Sum $\overline{1541}$ |

$67 \times 23 = 1{,}541$ **Product**

## 5.3 Other Multiplication Algorithms

2.    A            B
     29
     9, $R = 2$ and $2 \times 31 = 62$
     3, $R = 0$ and $0 \times 93 = 0$
     1, $R = 0$ and $0 \times 279 = 0$
     0, $R = 1$ and $1 \times 837 = 837$
                        Sum   899

     $29 \times 31 = 899$    **Product**

### Exercises I

Use the halve-double-sum technique to find each indicated product in Exercises 1 through 9. Check your answer by the standard multiplication algorithm.

1. $25 \times 82$.
2. $24 \times 38$.
3. $32 \times 33$.
4. $17 \times 91$.
5. $18 \times 84$.
6. $28 \times 61$.
7. $42 \times 51$.
8. $33 \times 43$.
9. $19 \times 43$.
10. Use the technique where 3 replaces 2 in the halve-double-sum technique to find each of the following products. (a) $47 \times 81$. (b) $42 \times 51$. (c) $25 \times 82$.

### Exercises II

Use the halve-double-sum technique to find each indicated product in Exercises 1 through 9. Check your answer by the standard multiplication algorithm.

1. $37 \times 84$.
2. $53 \times 82$.
3. $57 \times 38$.
4. $62 \times 111$.
5. $56 \times 213$.
6. $67 \times 334$.
7. $134 \times 243$.
8. $113 \times 241$.
9. $101 \times 312$.
10. Determine the technique necessary to find products where you divide by 4 and record the incomplete quotients. Then, find the following products by this method. Check using the halve-double-sum technique. (a) $134 \times 243$. (b) $113 \times 241$. (c) $101 \times 312$.

## 5.4 Supplementary Exercises

1. (a) Name the rational operations. (b) State the definition of division.

In each of Exercises 2 through 9, determine the number which $N$ represents.

2. $(45 + 3) \div N = 18 \div 3$.
3. $(N \div 4) - 3 = 2 \times 11$.
4. $(8 \times 3) \div N = 36 \div 18$.
5. $(180 \div 20) \times 3 = N - 28$.
6. $6 \times N = (28 \times 5) - 2$.
7. $235N = 19{,}270$.
8. $28(N + 3) = 2{,}856$.
9. $421N = 93{,}886$.
10. (a) Use the divisibility tests to show that 11 is a factor of 481,987.
    (b) Find the natural number $x$ such that $11x = 481{,}987$.
11. (a) Use the divisibility tests to show that 12 is a factor of 25,572.
    (b) Find the natural number $x$ such that $12x = 25{,}572$.
12. (a) Use the divisibility tests to show that 18 is a factor of 154,674.
    (b) Find the natural number $x$ such that $18x = 154{,}674$.
13. (a) Use the divisibility tests to show that 15 is a factor of 28,387,005.
    (b) Find the natural number $x$ such that $15x = 28{,}387{,}005$.
14. (a) Use the divisibility tests to show that 40 is a factor of 11,780,680.
    (b) Find the natural number $x$ such that $40x = 11{,}780{,}680$.

In Exercises 15 through 17, find the product by the halve-double-sum technique and check your answer by the standard algorithm.

15. $26 \times 271$.   16. $37 \times 83$.   17. $63 \times 97$.

In Exercises 18 through 20, find the product where 3 replaces 2 in the halve-double-sum technique.

18. $26 \times 271$.   19. $37 \times 83$.   20. $63 \times 97$.

# 6

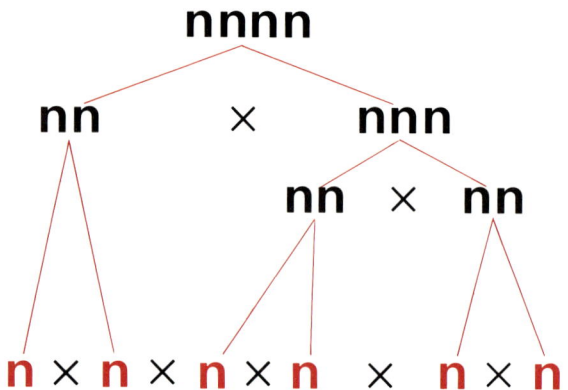

# FUNDAMENTAL THEOREM OF ARITHMETIC

## 6.1 Primes

We observed earlier that any natural number other than 1 has at least two different factors—itself and 1. Some natural numbers, such as 3, 5, 7, 11, 13, 17, etc., have *exactly two* factors. Any natural number which has exactly two distinct natural numbers as factors is called a prime number.

Many elementary questions about primes (or the set of primes) are difficult, if not impossible, to answer. Consider the following questions.

1. What are the first twenty-five primes?
2. For any given natural number $n$, is there a formula which will give the first $n$ primes?
3. For any given natural number $n$, how many primes are there less than $n$?
4. Is the set of prime numbers a finite set? (Is there a last prime?)
5. If a number is not a prime, can it be expressed as the product of primes? If the answer is "yes," can the number be expressed as the product of primes in more than one way?

We do not know the answer to Question 2, but we can give a *routine procedure* which will yield all primes less than, say, 100. First write all the natural numbers from 1 to 100 as in Figure 1. (This is tedious but not technically difficult.) Since 1 is not a prime, we cross it out. Since 2 has exactly two factors, it is a prime. Now, each multiple of 2 which is greater than 2 is not a prime since it will have 2 as a factor in addition to the factors of 1 and itself. Therefore, we cross out every second number in the list after 2. The next

|  |  |  |  |  |  |  |  |  |  |
|---|---|---|---|---|---|---|---|---|---|
| ~~1~~ | 2 | 3 | ~~4~~ | 5 | ~~6~~ | 7 | ~~8~~ | ~~9~~ | ~~10~~ |
| 11 | ~~12~~ | 13 | ~~14~~ | ~~15~~ | ~~16~~ | 17 | ~~18~~ | 19 | ~~20~~ |
| ~~21~~ | ~~22~~ | 23 | ~~24~~ | ~~25~~ | ~~26~~ | ~~27~~ | ~~28~~ | 29 | ~~30~~ |
| 31 | ~~32~~ | ~~33~~ | ~~34~~ | ~~35~~ | ~~36~~ | 37 | ~~38~~ | ~~39~~ | ~~40~~ |
| 41 | ~~42~~ | 43 | ~~44~~ | ~~45~~ | ~~46~~ | 47 | ~~48~~ | ~~49~~ | ~~50~~ |
| ~~51~~ | ~~52~~ | 53 | ~~54~~ | ~~55~~ | ~~56~~ | ~~57~~ | ~~58~~ | 59 | ~~60~~ |
| 61 | ~~62~~ | ~~63~~ | ~~64~~ | ~~65~~ | ~~66~~ | 67 | ~~68~~ | ~~69~~ | ~~70~~ |
| 71 | ~~72~~ | 73 | ~~74~~ | ~~75~~ | ~~76~~ | ~~77~~ | ~~78~~ | 79 | ~~80~~ |
| ~~81~~ | ~~82~~ | 83 | ~~84~~ | ~~85~~ | ~~86~~ | ~~87~~ | ~~88~~ | 89 | ~~90~~ |
| ~~91~~ | ~~92~~ | ~~93~~ | ~~94~~ | ~~95~~ | ~~96~~ | 97 | ~~98~~ | ~~99~~ | ~~100~~ |

Figure 1. *Sieve of Eratosthenes for primes less than 100.*

number in the list which is greater than 2 and has not been crossed out is 3; it is a prime. We cross out every third number after 3 since each of these is a multiple of 3 and not a prime. The next number in the list which is greater than 3 and has not been crossed out is 5; it is a prime. We cross out every fifth number after 5 since each of these is a multiple of 5 and not a prime. Occasionally, some numbers are crossed out more than once by this process but this makes no difference; the numbers which remain are primes. This method for finding prime numbers is referred to as the Sieve of Eratosthenes.

Not only is there no known formula to give all the primes but there is no known formula that will give us the number of primes less than any natural number. Of course, if we are given a specific number such as 200, the Sieve of Erastosthenes could be used to find all the primes less than 200 and also the number of primes less than 200.

Let us look at the concept of prime number in terms of sets. Consider a finite set with $a$ elements. The natural number $a$ is a prime if and only if the only way the set can be split into disjoint subsets with each containing exactly the same number of elements is by separating the set into disjoint sets where each set has exactly one element.

Any natural number different from 1 which is not a prime is called a composite number. Hence, the set of natural numbers is the union of three disjoint sets: the set of primes, the set of composite numbers, and the set containing just the number 1.

Let us answer two related questions concerning prime numbers

## 6.1 Primes

and composite numbers. (1) Is the set of prime numbers a finite set? (2) Can one find any given number of consecutive composite numbers? After observing the sieve technique for finding primes, one might conjecture correctly that the answer to Question (2) is "yes." This is not very difficult to verify. For example, consider the following nine consecutive numbers where $N = 10 \times 9 \times 8 \times 7 \times 6 \times 5 \times 4 \times 3 \times 2$:

$$N + 2, \quad N + 3, \quad N + 4,$$
$$N + 5, \quad N + 6, \quad N + 7,$$
$$N + 8, \quad N + 9, \quad N + 10.$$

Since 2 is a factor of 2 and since 2 is a factor of $N$, 2 is a factor of the sum $N + 2$; therefore, $N + 2$ is not a prime. Similarly, $N + 3$ is divisible by 3, $N + 4$ is divisible by 4, $N + 5$ is divisible by 5, etc. Hence, we have listed nine consecutive composite numbers. It should be obvious that this technique could be used to find any number of consecutive composite numbers. For example, if

$$M = 100 \times 99 \times 98 \times \cdots \times 4 \times 3 \times 2,$$

then $M + 2, M + 3, M + 4, \ldots, M + 99, M + 100$ are ninety-nine consecutive composite numbers.

The fact that we could find a million consecutive composite numbers might lead us to conjecture (incorrectly) that there was a last prime number and that the set of primes is a finite set. There is no last prime and the proof of this fact was known to Euclid over two thousand years ago. The proof, which is given in Chapter 12, is not very difficult but it does involve some logical subtleties which we prefer not to discuss at this time.

Since every natural number has 1 as a factor, *every pair of natural numbers has at least one common factor*. Some pairs of numbers, such as 14 and 15, do not have any factor in common except 1. This property for certain pairs of natural numbers is important enough to deserve a distinguishing name.

> Two natural numbers are called relatively prime if and only if 1 is their only common factor.

For example, 14 and 15 are relatively prime, and 3 and 5 are relatively prime. Since 6 and 14 have a factor in common other than 1, they are not relatively prime.

## Exercises I

1. (a) Use the Sieve of Eratosthenes to list all primes less than 200.
   (b) How many primes are there less than 200?
2. Why must every prime greater than 3 differ from the next by at least 2?
3. List all the pairs of primes less than 200 whose difference is 2.
4. List twelve consecutive composite numbers.

In Exercises 5 through 8 state whether or not the given numbers are relatively prime.

5. (a) 33 and 45.   (b) 36 and 37.
6. (a) 14 and 45.   (b) 23 and 38.
7. (a) 19 and 21.   (b) 18 and 27.
8. (a) 68 and 73.   (b) 44 and 55.
9. It is unknown whether or not the set of primes of the form $(N \times N) + 1$ where $N$ is a natural number is a finite set. Find five primes of the form $(N \times N) + 1$.
10. Try to express every even number greater than 4 and less than 101 as the sum of two (not necessarily different) primes.
    (a) Can it be done?
    (b) Can any of these even numbers be expressed as the sum of two primes in more than one way?

## Exercises II

1. Let $p$ and $q$ be two different prime numbers. Are they relatively prime?
2. Let $p$ and $q$ be two different prime numbers. Use the definition of $s(n)$ on page 61 to answer the following. (a) What is $s(p)$? (b) What is $s(q)$? (c) What is $s(pq)$?
3. Let $p$ and $q$ be two different prime numbers. Use the definition of $d(n)$ on page 61 to answer the following. (a) What is $d(p)$? (b) What is $d(q)$? (c) What is $d(pq)$?
4. Use previously worked exercises and whatever additional examples are necessary to conjecture the conditions under

which $s(m) \times s(n) = s(mn)$ where $m$ and $n$ are natural numbers.
5. Use previously worked exercises and whatever additional examples are necessary to conjecture the conditions under which $d(m) \times d(n) = d(mn)$ where $m$ and $n$ are natural numbers.

Let $\phi(n)$ be the number of natural numbers less than $n$ which are relatively prime to $n$. For example $\phi(6) = 2$ since 1 and 5 are the only two numbers less than 6 which are relatively prime to 6. Find each number indicated in Exercises 6 through 15.

6. $\phi(3)$.    7. $\phi(5)$.    8. $\phi(7)$.    9. $\phi(15)$.    10. $\phi(12)$.
11. $\phi(28)$.    12. $\phi(35)$.    13. $\phi(40)$.    14. $\phi(45)$.    15. $\phi(60)$.
16. Let $p$ be a prime. What is $\phi(p)$?
17. Is $\phi(5) \times \phi(7) = \phi(35)$?
18. Is $\phi(3) \times \phi(15) = \phi(45)$?

## 6.2 Fundamental Theorem of Arithmetic

Consider the number 1890. Since the sum $1 + 8 + 9 + 0 = 18$ has 9 as a factor, the divisibility test for 9 ensures that 1890 has 9 as a factor. Using the division algorithm, we find that $1890 \div 9 = 210$; thus,

$$1890 = 9 \times 210.$$

From the divisibility test for 10, we conclude that 10 is also a factor of 1890. Since $1890 \div 10 = 189$,

$$1890 = 10 \times 189.$$

Thus, 1890 has been expressed as a product in two different ways. Since $9 = 3 \times 3$ and $210 = 21 \times 10 = 7 \times 3 \times 5 \times 2$, we see that

$$1890 = 3 \times 3 \times 7 \times 3 \times 5 \times 2.$$

Here, we have expressed 1890 as a product of primes. Sometimes, these prime factors can be found rather easily using what is called a *factor tree*. The factor tree associated with this prime decomposition of 1890 is as follows.

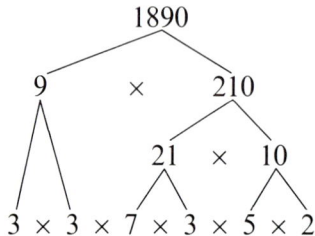

Another factor tree is obtained by starting with the fact that $1890 = 10 \times 189$.

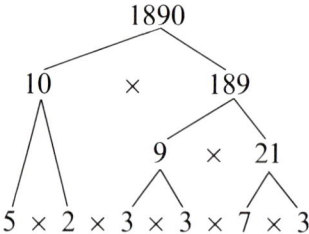

Although the factor trees are different, it is important to observe that the "roots" (base numbers) are the same. In other words, each of the decompositions of 1890 into the product of primes is the same, except for the order in which the primes are written.

It is an important fact that any composite number can be expressed as the product of primes in one and only one way, except for the order in which the primes are listed. This statement is called the Fundamental Theorem of Arithmetic. The Fundamental Theorem of Arithmetic states a fact which is rather easy to understand (and believe), but the proof of this theorem is not quite as simple. For this reason we defer a proof until Chapter 12. However, the examples and exercises should convince us that the Fundamental Theorem of Arithmetic is a true statement.

### Exercises I

Use the divisibility tests, the division algorithm, or a factor tree to express each of the following numbers as a product of primes.

1. 360.
2. 1,320.
3. 266.
4. 480.
5. 1,240.
6. 225.
7. 6,946.
8. 204.
9. 9,360.
10. 1,089.
11. 720.
12. 1,444.

6.3 G.C.F. and L.C.M.

### Exercises II

1. If 1 were included in the set of prime numbers would the Fundamental Theorem of Arithmetic be a true statement?
2. If 1 were included in the set of composite numbers, would the Fundamental Theorem of Arithmetic be a true statement?

In Exercises 3 through 10, use the divisibility tests, the division algorithm, or a factor tree to express each of the following numbers as a product of primes.

| | | | |
|---|---|---|---|
| 3. 475. | 4. 870. | 5. 1,275. | 6. 2,868. |
| 7. 23,248. | 8. 47,412. | 9. 6,080. | 10. 30,448. |

## 6.3 G.C.F. and L.C.M.

Let $S$ be the set of factors of 20 and let $T$ be the set of factors of 30. Thus, $S = \{1, 2, 3, 4, 5, 10, 20\}$ and $T = \{1, 2, 3, 4, 5, 10, 15, 30\}$. The set $S \cap T = \{1, 2, 5, 10\}$ is the set of *common factors* of 20 and 30. We can see that 10 is the greatest element in $S \cap T$; thus 10 is the greatest common factor of 20 and 30.

Let $a$ and $b$ be any natural numbers and let $S$ be the set of factors of $a$ and let $T$ be the set of factors of $b$. We know that $1 \in S$ and $1 \in T$; thus, $S \cap T$ is not empty. Since both $S$ and $T$ have greatest elements (namely, $a$ and $b$, respectively), $S \cap T$ has a greatest element. The greatest element in $S \cap T$ is called the greatest common factor of $a$ and $b$ and is often denoted by G.C.F.

To find the G.C.F. of two natural numbers $a$ and $b$ is quite routine if we have found all the factors of $a$ and $b$. However, if $a$ and $b$ are very large, then finding and listing all the factors of each number could be a difficult task.

Another way to find the G.C.F. of two numbers such as 20 and 30 is exhibited as follows. Express each number as the product of primes.

$$20 = 2 \times 2 \times 5$$

and

$$30 = 2 \times 3 \times 5.$$

Observe that each number has 2 as a common factor and each has 5 as a common factor. Since there are no other common factors, except 1, 2 × 5 = 10 is the G.C.F. of the given numbers. Of course, a factor tree could have been used to express each number as the product of primes.

*Examples*

1. Find the G.C.F. of 165 and 630.

    **Method I.** The set of factors of 165 is

    $$S = \{1, 3, 5, 11, 15, 33, 55, 165\}.$$

    The set of factors of 630 is

    $$T = \{1, 2, 3, 5, 6, 7, 9, 10, 14, 15, 18, 21, 30, 35,$$
    $$42, 45, 63, 70, 90, 105, 126, 210, 315, 630\}.$$

    Thus,

    $$S \cap T = \{1, 3, 5, 15\},$$

    and the G.C.F. is 15.

    **Method II.**

    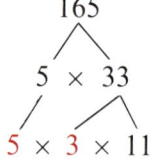
    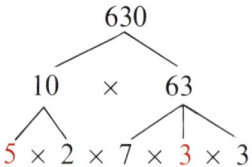

    The G.C.F. is 3 × 5 = 15.

2. Find the G.C.F. of 1460 and 2190. (Note: Before continuing, if the student would list the set of factors of each of these numbers without looking at the list below, the advantages of Method II would be quite obvious.)

    **Method I.** The set of factors of 1460 is

    $$S = \{1, 2, 4, 5, 10, 20, 73, 146, 292, 365, 730, 1460\}.$$

    The set of factors of 2190 is

    $$T = \{1, 2, 3, 5, 6, 10, 15, 30, 73, 146, 219,$$
    $$365, 438, 730, 1095, 2190\}.$$

    Thus,

    $$S \cap T = \{1, 2, 5, 10, 73, 146, 365, 730\},$$

    and the G.C.F. is 730.

## 6.3 G.C.F. and L.C.M.

**Method II.**

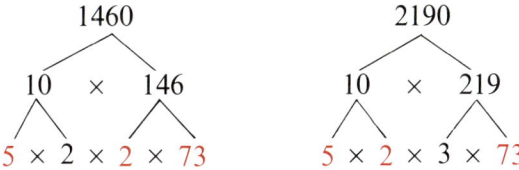

The G.C.F. is $5 \times 2 \times 73 = 730$.

Of course, expressing a number as a product of primes can be a a tedious and difficult task, but Method II is more efficient to find the G.C.F. than by listing all the factors of the numbers. In the next section, we shall discuss a routine procedure to find the G.C.F. for which it is unnecessary to find the factors of each number or to express each number as a product of primes.

Let $S$ be the set of multiples of 20 and let $T$ be the set of multiples of 30. Thus,

$$S = \{20, 40, 60, 80, 100, 120, 140, 160, 180, \ldots\}$$

and

$$T = \{30, 60, 90, 120, 150, 180, 210, 240, 270, \ldots\},$$

where $S$ and $T$ are infinite sets. Since $20 \times 30 = 600$ is a multiple of both 20 and 30, the set $S \cap T$ is a set of natural numbers which is not empty. By the well-ordering property, we know that $S \cap T$ contains a least element. Since

$$S \cap T = \{60, 120, 180, 240, \ldots\},$$

the least element is 60; thus, 60 is the *least* element in the set of *common multiples* of 20 and 30.

Let $a$ and $b$ be any natural numbers. Let $S$ be the set of multiples of $a$ and let $T$ be the set of multiples of $b$. Since $ab$ is a multiple of both $a$ and $b$, $S \cap T$ is not empty and by the well-ordering property $S \cap T$ contains a least element. The least element in the set of common multiples is called the least common multiple of $a$ and $b$ and it is often denoted by L.C.M.

### Examples

1. Find the L.C.M. of 30 and 42.

   *Solution.* The set of multiples of 30 is

$S = \{30, 60, 90, 120, 150, 180, 210, 240, 270,$
$$300, 330, 360, 390, 420, 450, \ldots\}.$$

The set of multiples of 42 is

$T = \{42, 84, 126, 168, 210, 252, 294, 336,$
$$378, 420, 462, \ldots\}.$$

Thus,
$$S \cap T = \{210, 420, 630, \ldots\},$$
and the L.C.M. is 210.

2. Find the L.C.M. of 18 and 64.

*Solution.* The set of multiples of 18 is

$S = \{18, 36, 54, 72, 90, 108, 126, 144, 162, 180,$
$$198, 216, 234, 252, \ldots\}.$$

The set of multiples of 63 is
$$T = \{63, 126, 189, 252, 315, \ldots\}.$$

Thus,
$$S \cap T = \{126, 252, 378, \ldots\},$$
and the L.C.M. is 126.

We have exhibited one technique to find the L.C.M. of two natural numbers. In several earlier exercises, as well as in the examples of this section, we have seen for numbers $a$ and $b$ where $G$ is the G.C.F. and $L$ is the L.C.M. that

$$L \times G = a \times b.$$

As another example, consider the two natural numbers $a = 12$ and $b = 28$. The greatest common factor of the two numbers is $G = 4$. Notice that $Gx = (4)(3) = 12 = a$ and $Gy = (4)(7) = 28 = b$, and that $x = 3$ and $y = 7$ are relatively prime numbers. Furthermore, $Gxy = (4)(3)(7)$ is a multiple of 12 and a multiple of 28. In fact, since 3 and 7 are relatively prime, $Gxy = (4)(3)(7)$ is the least common multiple $L$ of $a = 12$ and $b = 28$. Observe that

$$\begin{aligned} L \times G &= (Gxy) \times G \\ &= (Gx) \times (Gy) \\ &= a \times b \end{aligned}$$

## 6.3 G.C.F. and L.C.M.

If $L \times G = a \times b$, then $L = (a \times b) \div G$. This, of course, gives us another technique, and usually a more efficient method, to find the L.C.M. of $a$ and $b$ *if* the G.C.F. is known.

### Examples

1. Find the L.C.M. of 165 and 630.

   *Solution.* In Example 1, page 90, we found that the G.C.F. is 15. Hence,
   
   $$\text{L.C.M.} = (165 \times 630) \div 15 = 6{,}930.$$

2. Find the L.C.M. of 1,460 and 2,190.

   *Solution.* In Example 2, page 90, we found that the G.C.F. is 730. Hence,
   
   $$\text{L.C.M.} = (1{,}460 \times 2{,}190) \div 730 = 4{,}380.$$

Now, let us generalize from our examples to show that $L = (a \times b) \div G$ for any two natural numbers $a$ and $b$. First, since the greatest common factor $G$ is a factor of both $a$ and $b$, there exist natural numbers $x$ and $y$ such that $Gx = a$ and $Gy = b$. Since $G$ is the greatest common factor of $a$ and $b$, both $x$ and $y$ have no other factor in common except 1; that is, $x$ and $y$ are relatively prime. Now, $a \times b = (Gx) \times (Gy) = (Gxy) \times G$. From $Gx = a$ and $Gy = b$, we see that $Gxy = ay$ and $Gxy = bx$; thus, $Gxy$ is a multiple of both $a$ and $b$. Also, since $x$ and $y$ have no common factor except 1, $Gxy$ is the least common multiple of $a$ and $b$; thus, $L = Gxy$. Consequently, $a \times b = L \times G$ and $L = (a \times b) \div G$.

### Exercises I

Find the G.C.F. and the L.C.M. of each pair of numbers in Exercises 1 through 9.

1. 78 and 102.
2. 105 and 270.
3. 23 and 37.
4. 48 and 57.
5. 11 and 33.
6. 36 and 101.
7. 690 and 810.
8. 1,206 and 1,260.
9. 6,600 and 11,781.

10. (a) Find the G.C.F. of 154, 210, and 324; that is, find the greatest number in the intersection of the sets of factors of the three numbers.
    (b) Find the L.C.M. of 154, 210, and 324.
    (c) For two numbers $a$ and $b$ we found that $L \times G = a \times b$. Is it true for three numbers $a$, $b$, and $c$ that $L \times G = a \times b \times c$?

### Exercises II

Find the G.C.F. and the L.C.M. for each pair of numbers in Exercises 1 through 12.

1. 105 and 140.
2. 31 and 48.
3. 68 and 71.
4. 220 and 385.
5. 41 and 43.
6. 410 and 205.
7. 95 and 144.
8. 92 and 115.
9. 145 and 232.
10. 451 and 820.
11. 403 and 2,170.
12. 308 and 385.

## 6.4 Euclidean Algorithm

Suppose we want to find the G.C.F. of 4,981 and 10,608. If it were easy to find the prime factors of each number, then we could use the technique described in Section 6.3 to find the G.C.F. However, if the prime factors cannot be easily found, then we should seek another method. One can be obtained and it is described as follows.

Consider the division (with remainder) of 10,608 by 4,981.

$$\textbf{divisor} = 4{,}9\,8\,1 \overline{\smash{\big)}\,\begin{array}{r} 2 \\ 1\,0{,}6\,0\,8 \\ 9\,9\,6\,2 \\ \hline 6\,4\,6 \end{array}} = \textbf{remainder}$$

The incomplete quotient is 2 and the remainder is 646. Thus,

$$10{,}608 = 2(4{,}981) + 646.$$

Now, *any* factor of the divisor 4,981 and the remainder 646 is also a factor of 10,608 since if two numbers have a common factor then so does their sum. Since

$$646 = 10{,}608 - 2(4{,}981),$$

## 6.4 Euclidean Algorithm

*any* factor of the given numbers is also a factor of 646 since if two numbers have a common factor then so does their difference. From this we can conclude that the *greatest common factor of the two given numbers is also the greatest common factor of the remainder* 646 *and the divisor* 4,981.

Now, let us repeat the division process and find the incomplete quotient and remainder in the division of 4,981 by 646

$$\mathbf{divisor} = 646 \overline{)\begin{array}{r} 7 \\ 4{,}981 \\ 4522 \\ \hline 459 \end{array}} = \mathbf{remainder}$$

From this, we can conclude as we did above that the greatest common factor of 646 and 4,981 is the greatest common factor of the remainder 459 and the divisor 646. Since the G.C.F. of 646 and 4,981 is the G.C.F. of the original numbers 4,981 and 10,608, the G.C.F. of 459 and 646 is the G.C.F. of 4,981 and 10,608.

Continuing, we find from

$$\mathbf{divisor} = 459 \overline{)\begin{array}{r} 1 \\ 646 \\ 459 \\ \hline 187 \end{array}} = \mathbf{remainder}$$

that the G.C.F. of 187 and 459 is the G.C.F. of the original numbers 4,981 and 10,608. Again, from

$$\mathbf{divisor} = 187 \overline{)\begin{array}{r} 2 \\ 459 \\ 374 \\ \hline 85 \end{array}} = \mathbf{remainder}$$

we conclude that the G.C.F. of 85 and 187 is the G.C.F. of 4,981 and 10,608. Let us continue this process.

$$\mathbf{divisor} = 85 \overline{)\begin{array}{r} 5 \\ 459 \\ 425 \\ \hline 34 \end{array}} = \mathbf{remainder}$$

The G.C.F. of 34 and 85 is the G.C.F. of 4,981 and 10,608.

$$\text{divisor} = 34 \overline{\smash{\big)}85} \phantom{0} 2$$
$$\underline{68}$$
$$17 = \text{remainder}$$

The G.C.F. of 17 and 34 is the G.C.F. of 4,981 and 10,608.

$$\text{divisor} = 17 \overline{\smash{\big)}34} \phantom{0} 2$$
$$\underline{34}$$
$$0 = \text{remainder}$$

Since 17 is a *factor* of 34, the G.C.F. of these two numbers is 17. Thus, 17 is the G.C.F. of 17 and 34 which implies that 17 is the G.C.F. of 4,981 and 10,608.

The preceding argument is a little long, but the student should make every effort to understand why the process does indeed give the G.C.F. of the two given numbers. Also, the process may seem rather cumbersome, but it can be shortened considerably. The following display exhibits a routine procedure to perform the successive divisions necessary to find the G.C.F. of 4,981 and 10,608.

```
          2
4,981 ) 10,608
        9962           7
         646 ) 4,981
               4522         1
                459 ) 646
                      459         2
                      187 ) 459
                            374         5
                             85 ) 459
                                  425         2
                                   34 ) 85
                                        68         2
                         G.C.F. =       17 ) 34
                                              34
                                               0
```

This process of dividing the smaller of two numbers into the larger and then successively dividing each remainder into the preceding divisor to find the G.C.F. of the given numbers is called the Euclidean algorithm. The last non-zero remainder in the Euclidean

## 6.4 Euclidean Algorithm

algorithm is the G.C.F. of the two given numbers; a detailed proof of this fact is given in Chapter 12.

In the Euclidean algorithm, since each remainder need only be less than the divisor, it could appear that if the original divisor is a number as large as 4,981 then the number of repeated divisions required to find the G.C.F. might be prohibitive. Actually, we can prove a theorem called *Lame's Theorem*\* which states that the number of repeated divisions necessary to get a zero remainder cannot exceed 5 times the number of digits in the original divisor. In our example, this means that the number of divisions cannot exceed $5 \times 4 = 20$. In practice, the number of divisions required is generally considerably less than 5 times the number of digits in the original divisor; it was 7 in the example.

### Example

Find the G.C.F. and L.C.M. of 11,040 and 49,151.

```
              4
11,040 | 49,151
         44160            2
         ‾‾‾‾‾     4,991 | 11,040
                          9982           4
                          ‾‾‾‾‾  1,058 | 4,991
                                        4232          1
                                        ‾‾‾‾‾   759 | 1,058
                                                      759         2
                                                      ‾‾‾  299 | 759
                                                                598         1
                                                                ‾‾‾  161 | 299
                                                                            161          1
                                                                            ‾‾‾  138 | 161
                                                                                        138         6
                                                          G.C.F. =            23 | 138
                                                                                        138
                                                                                        ‾‾‾
                                                                                          0
```

Thus, 23 is the G.C.F.; the L.C.M. is

$$L = (11{,}040 \times 49{,}151) \div 23 = 23{,}562{,}240.$$

---

\*For a proof, see *An Introduction to Mathematics* by B. K Youse (Boston: Allyn and Bacon, Inc., 1970).

### Exercises I

Use the Euclidean algorithm to find the G.C.F. of each of the pairs of numbers in Exercises 1 through 9.
    1. 1,288 and 1,176.        2. 2,520 and 3,591.
    3. 2,436 and 1,943.        4. 1,488 and 1,680.
    5. 2,680 and 2,881.        6. 2,610 and 4,466.
    7. 3,604 and 6,188.        8. 5,043 and 8,241.
    9. 9,324 and 10,545.

Find the L.C.M. of each pair of number in the indicated exercises.
    10. Exercise 1.        11. Exercise 2.        12. Exercise 3.
    13. Exercise 4.        14. Exercise 5.        15. Exercise 6.
    16. Exercise 7.        17. Exercise 8.        18. Exercise 9.

### 6.5 Supplementary Exercises

    1. List all pairs of primes less than 200 whose difference is 4.
    2. List eight consecutive composite numbers.

In Exercises 3 through 6, state whether or not the given numbers are relatively prime.
    3. (a) 362 and 471. (b) 844 and 926.
    4. (a) 125 and 357. (b) 681 and 232.
    5. (a) 89 and 98. (b) 123 and 321.
    6. (a) 978 and 565. (b) 868 and 1,235.

In Exercises 7 through 12, use a factor tree to express each number as the product of primes.
    7. 300.        8. 1492.        9. 1066.
    10. 1776.        11. 1630.        12. 1984.

In Exercises 13 through 18, find the G.C.F. of the two numbers.
    13. 78 and 36.        14. 54 and 824.        15. 882 and 1953.
    16. 47 and 79.        17. 570 and 646.        18. 1012 and 2728.
    19. Find the L.C.M. of each pair of numbers in Exercises 13 through 18.
    20. (a) Find the G.C.F. of 640, 1,036, and 1,776.
        (b) Find the L.C.M. of 640, 1,036, and 1,776.

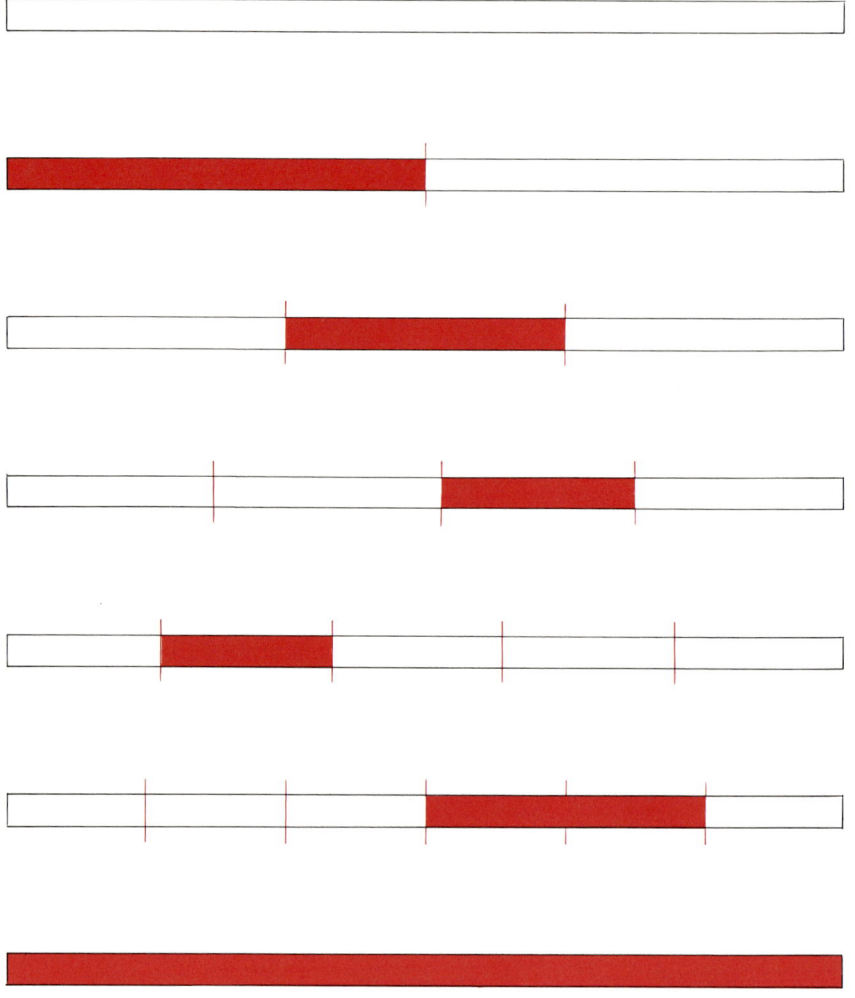

# THE RATIONAL NUMBERS

## 7.1 The Rational Numbers

Many years after the development of the counting numbers, particularly when man began to develop a commerce-based culture, it became necessary for him to establish units of weights and measure. After a unit of length is chosen, it becomes immediately evident that not all lengths are "measurable" in terms of the counting numbers; that is, a given length need not be equal to any specific counting number of units.

The early geometers knew not only how to bisect a line segment but also how to divide a line segment into any number of equal parts. In Figure 1, if $AB$ is a unit segment, then the darkened segment in (a) is 1 part of 2 equal parts. In (b), the darkened segment is 1 part of 3 equal parts, and in (c) it is 2 parts of 3 equal parts. We could continue to use the counting numbers in pairs to identify

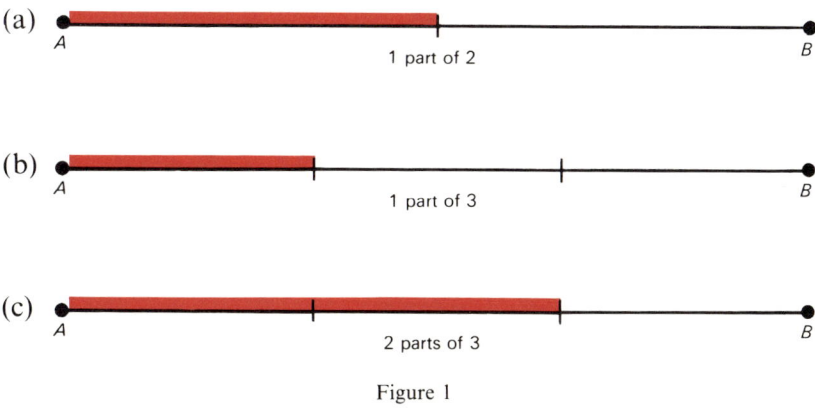

Figure 1

such lengths, but it would (and did for many years) hinder the development of a sound arithmetic. For man, the advancement to a stage where numbers were used for a purpose other than for counting was a towering achievement. This profound change, taking place over hundreds of years, necessitated a new synthesis of the number concept.

Today, we use the symbol "$\frac{1}{3}$" (read "one-third"), called a fraction to represent the length of the segment $AC$ in Figure 1b. Eventually, we make the inevitable abstraction and consider the number $\frac{1}{3}$ worthy of interest in its own right, but at the outset we think of it as representing one part of some quantity which has been separated into 3 equal parts.

A fraction is also employed to assign area to certain types of regions. In Figure 2, if the circular region in (a) is considered to have unit area, then the area of the shaded region in (b) is $\frac{1}{2}$ since it is 1 part of the 2 equal parts into which the entire circular region has been divided. We see that the shaded region in (c) would be $\frac{1}{4}$ of that in (a). Since the shaded region in (d) is 2 parts of 8 equal

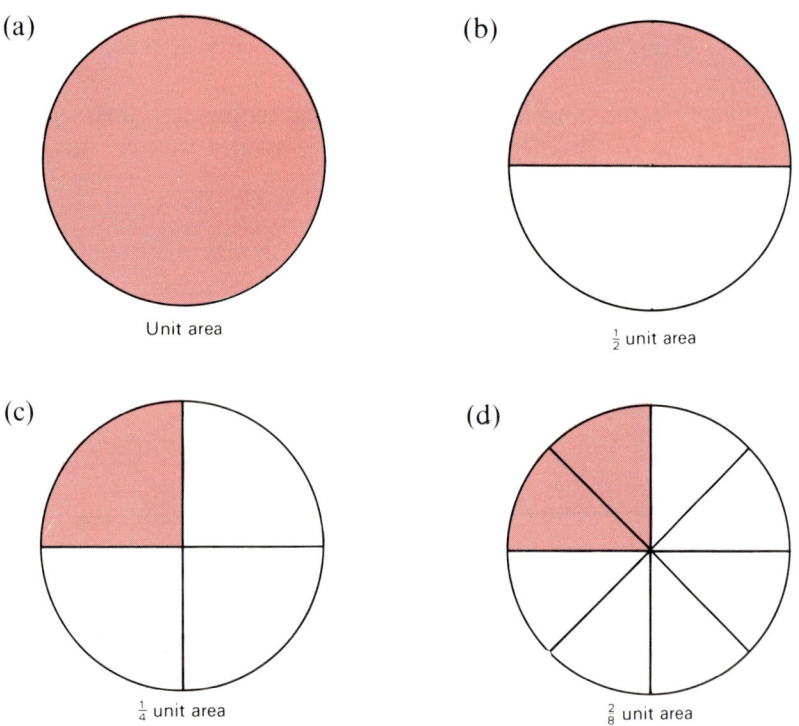

Figure 2

## 7.1 The Rational Numbers

parts, its area would be denoted by "$\frac{2}{8}$". Since the shaded regions in (c) and (d) have the same area, we see that $\frac{1}{4} = \frac{2}{8}$.

Closely connected with the idea of using numbers to measure is the idea of associating numbers with points on a line. Let us first discuss the method used to construct a correspondence between the whole numbers and certain points on a line.

The term *line* refers to a straight line which extends indefinitely in each direction. A *ray* is that part of a line which extends indefinitely in one direction from some point $P$ on the line. The point $P$ is called the initial point, or origin, of the ray. We begin to construct what we call a number line by marking off equal segments on a ray to the right of the origin as in Figure 3. We associate 0 with the origin $P$. Next, we associate 1 with the end point of the first segment to the right of $P$ and call this a unit length. The number 2 is paired with the end point of the second segment, the number 3 is paired with the end point of the third segment, etc. Thus, we have a method for pairing each whole number with a point on the given ray.

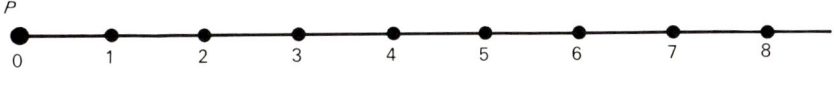

Figure 3. *Number line.*

We notice immediately that there are points on the number line which are not paired with any whole number. Since we use a line segment to denote lengths, we want eventually to have a number associated with every point on the ray. In other words, one of our goals is to fill up the number line with numbers.

As we said earlier, a simple geometric construction can be used to divide a line segment into any number of equal parts. Suppose we want to divide the unit segment between 0 and 1 into thirteen equal segments; we proceed as follows. Draw a line $PA$ at any convenient angle to the line $PB$ as in Figure 4. Then, mark off on $PA$ thirteen

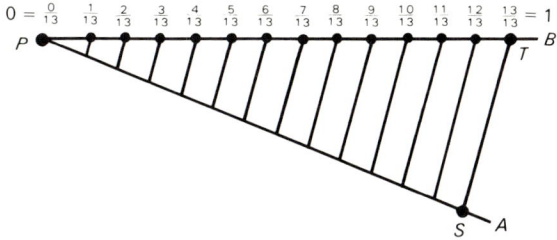

Figure 4. *Rational numbers.*

equal line segments of any convenient length. Next, draw a line from the end point of the thirteenth segment, *S*, to the unit point, *T*. If a line parallel to the line *ST* is constructed through each end point of the line segments on *PA*, then the unit interval *PT* will be cut by these parallel lines into thirteen equal segments. We call the end point of each such segment on the unit segment a rational point; the number representing the length from the origin to one of these rational points is called a rational number. The rational number associated with the end point of the first segment to the right of the origin is denoted by the fraction "$\frac{1}{13}$"; it represents the measure of 1 of 13 equal segments into which the unit length has been divided. The rational number associated with the end point of the second segment is denoted by "$\frac{2}{13}$"; the rational number associated with the end point of the third segment is denoted by "$\frac{3}{13}$"; etc. If we also divide the segment between 1 and 2 into 13 equal segments, then the rational number denoted by the fraction "$\frac{14}{13}$" would be associated with the end point of the first segment to the right of 1, the end point of the fourteenth such segment to the right of the origin. (See Figure 5.)

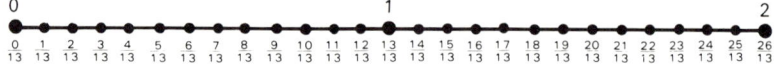

Figure 5. Rational numbers.

The reader should give particular attention to the distinction between the terms *rational number* and *fraction*. If the unit segment has been divided into 26 equal segments, then the *different* fractions "$\frac{4}{26}$" and "$\frac{2}{13}$" would represent the *same* rational number; thus, $\frac{4}{26} = \frac{2}{13}$.

In general, if *m* is a *whole number* and if *n* is a *natural number*, then the fraction "$\frac{m}{n}$" denotes a rational number. The fraction "$\frac{0}{n}$" is associated with the origin of the number line and "$\frac{m}{1}$" is associated with the same point as the whole number *m*. The number *m* is called the numerator of the fraction and the number *n* is called the denominator of the fraction. Geometrically, for the rational number denoted by the fraction "$\frac{m}{n}$," the denominator *n* denotes the number of equal segments into which the unit segment is divided. The number $\frac{m}{n}$ is associated with the end point of the *m*th such segment to the right of the origin.

A proper fraction is one where the numerator is a *natural* number

## 7.1 The Rational Numbers

less than the denominator. For example, "$\frac{1}{2}$," "$\frac{3}{5}$," and "$\frac{6}{19}$" are proper fractions. An **improper fraction** is one in which the numerator is greater than or equal to the denominator. For example, "$\frac{4}{3}$," "$\frac{23}{19}$," and "$\frac{13}{13}$" are improper fractions. Geometrically, proper fractions represent rational points between 0 and 1 on the number line. All rationals greater than or equal to 1 can be represented by improper fractions. Notice that "$\frac{0}{6}$" is not considered to be a proper or an improper fraction. Not putting the fraction "$\frac{0}{n}$" into either the set of proper or set of improper fractions is done for convenience; it just simplifies the development of the rationals.

### Exercises I

In Exercises 1 through 4, determine a fractional representation for the number associated with the given length.

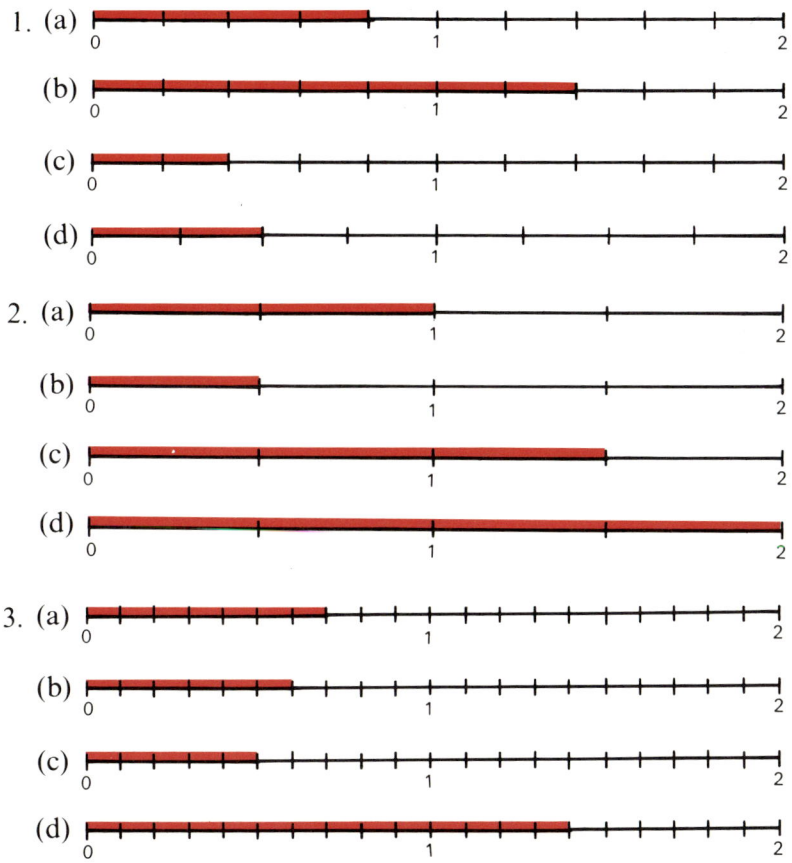

106   The Rational Numbers

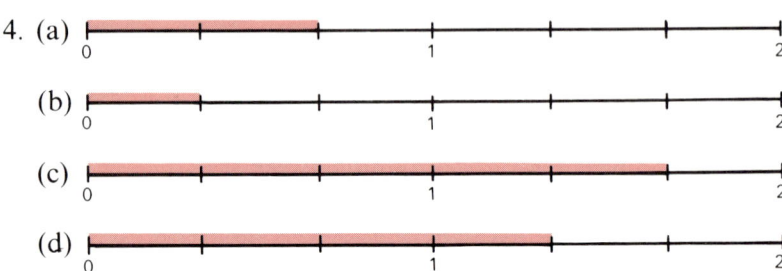

In Exercises 5 through 8, if each circular region has unit area determine a fractional representation for the rational number associated with the shaded region of each circular region.

5. (a)   (b)

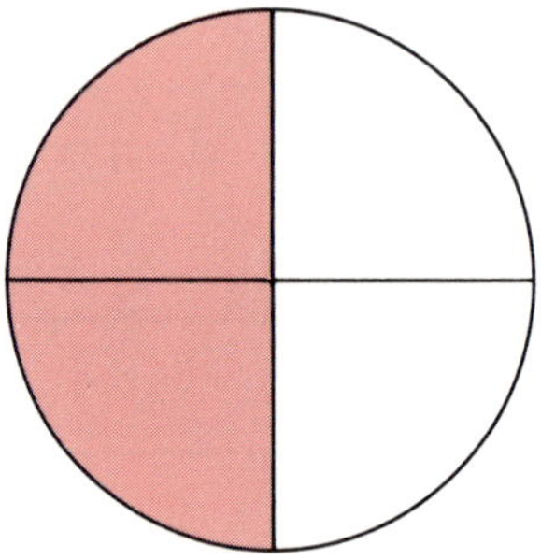

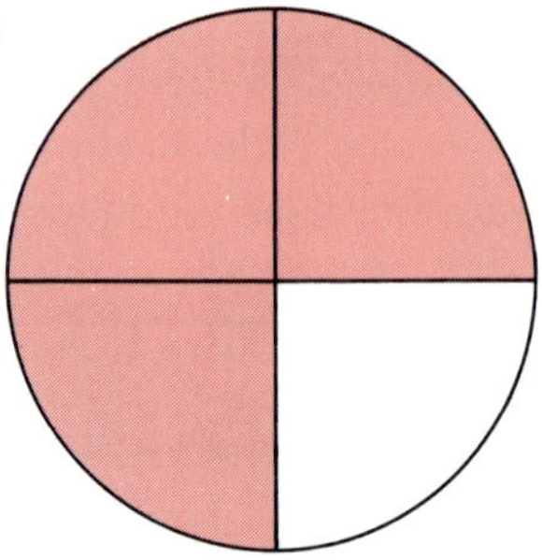

6. (a) 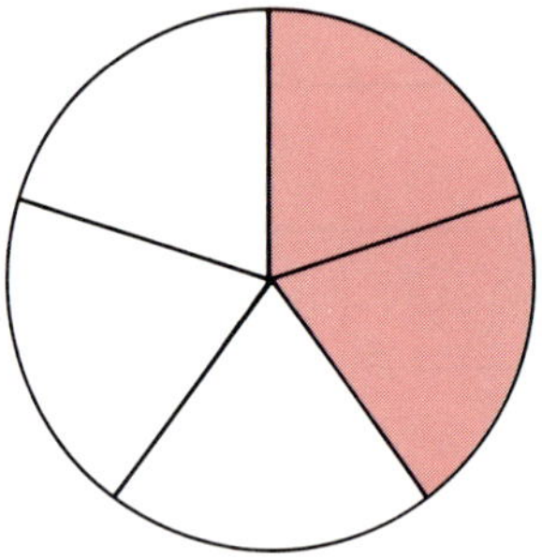   (b)

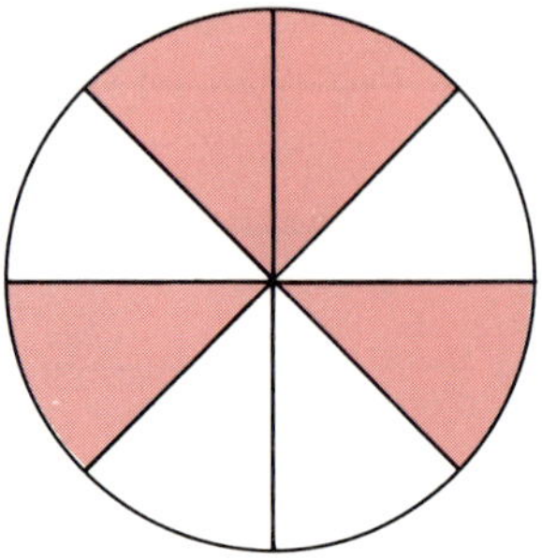

7. (a) 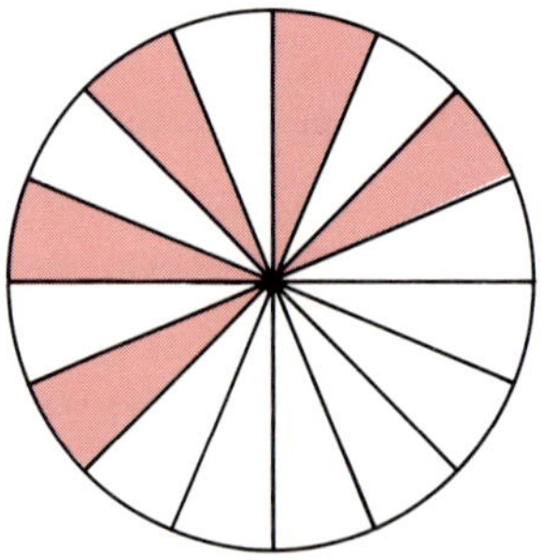   (b)

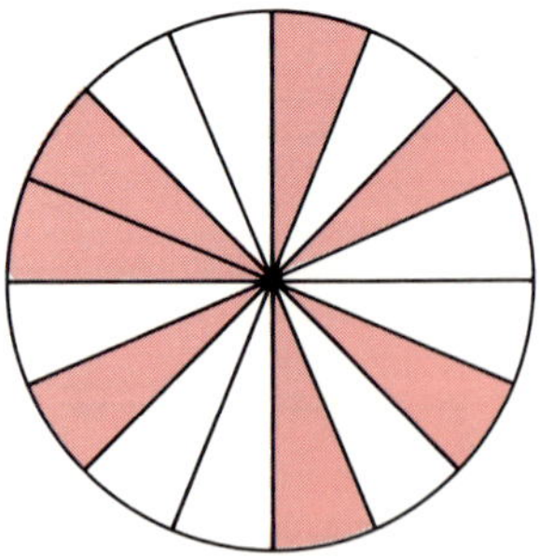

7.1 The Rational Numbers

8. (a)    (b)

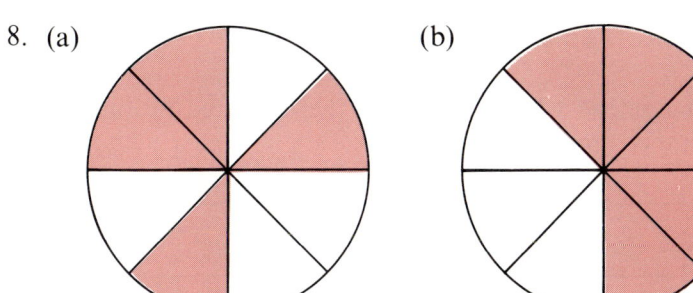

9. Give at least two fractional representations for the number associated with each indicated rational point.

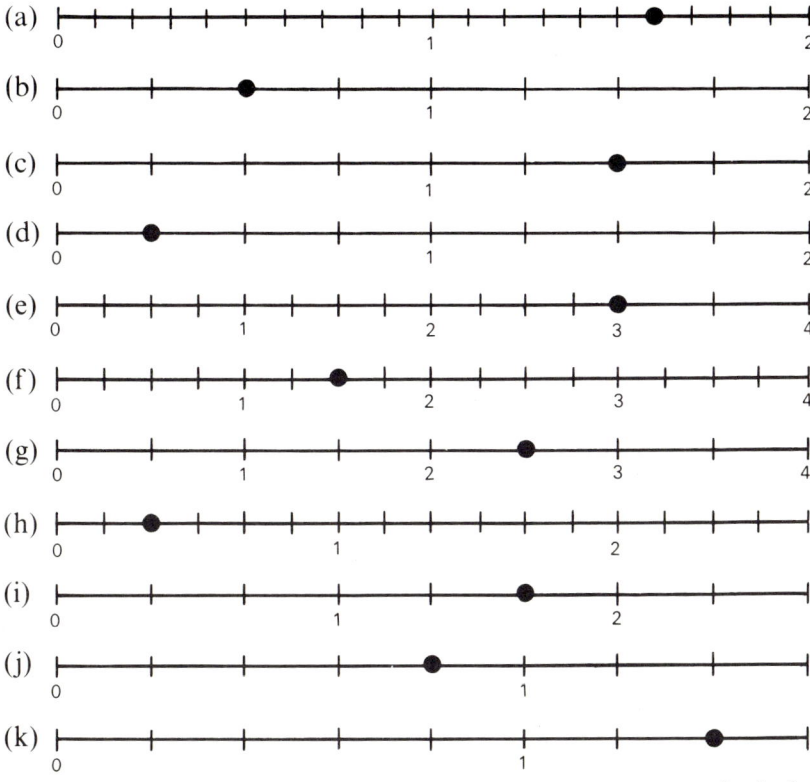

10. (a) Which of the following are proper fractions? $\frac{8}{9}, \frac{9}{4}, \frac{2}{6}, \frac{19}{3}, \frac{0}{7}, \frac{11}{11}, \frac{3}{1}, \frac{1}{12}, \frac{5}{7}, \frac{18}{3}, \frac{14}{22}, \frac{13}{19}.$
    (b) In part (a), which are improper fractions?
    (c) List each proper fraction in part (a) whose numerator is an odd number.
    (d) List the fractions in part (a) whose numerator and denominator are relatively prime natural numbers.

## Exercises II

Give a fraction which represents the indicated rational number on the number line.

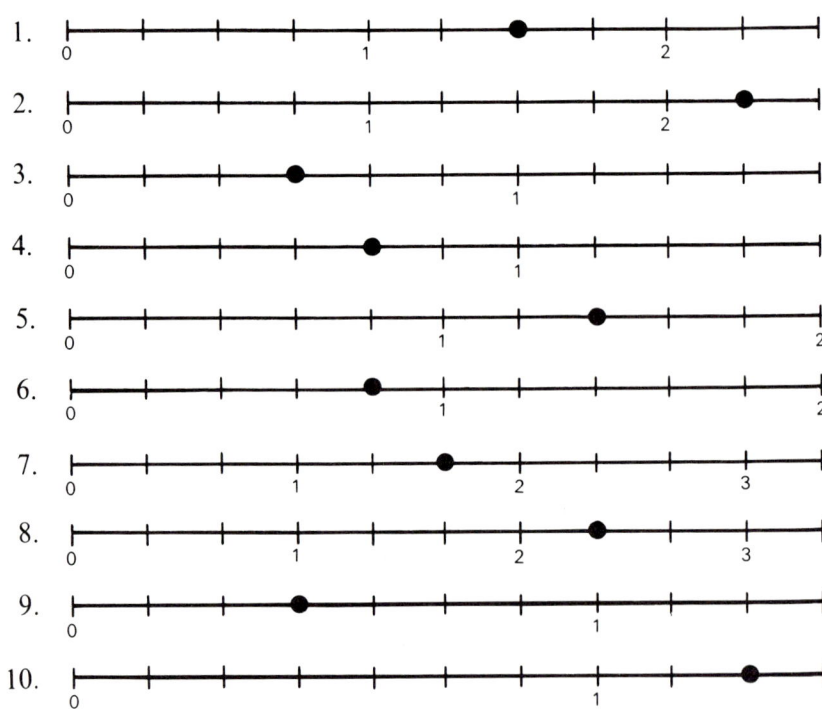

## 7.2 Equality and Inequality

Let $\frac{m}{n}$ be a rational number. From our observations in the last section, we conclude that if $p$ is a natural number then $\frac{rm}{pn} = \frac{m}{n}$. For example,

$$\frac{2}{3} = \frac{2 \cdot 2}{2 \cdot 3} = \frac{4}{6} \quad \text{and} \quad \frac{2}{3} = \frac{4 \cdot 2}{4 \cdot 3} = \frac{8}{12}.$$

Thus, we see that there are infinitely many fractional representations (names) for each rational number. Furthermore, we should observe that $\frac{2}{3} = \frac{4}{6}$ implies that $2 \times 6 = 3 \times 4$. In general, two rational numbers

$$\frac{x}{y} \quad \text{and} \quad \frac{u}{v} \quad \text{are equal if and only if } xv = yu.$$

## 7.2 Equality and Inequality

For two natural numbers $m$ and $n$, let $G$ be the greatest common factor of $m$ and $n$. For example, if $m = 8$ and $n = 20$ then $G = 4$. Thus, $m = Gx$ and $n = Gy$ where $x$ and $y$ are relatively prime. In our example, $8 = 4 \cdot 2$ and $20 = 4 \cdot 5$ where 2 and 5 are relatively prime. Since

$$\frac{m}{n} = \frac{Gx}{Gy} = \cdot \frac{x}{y},$$

any rational number different from zero can be written in fractional notation where the numerator and denominator are relatively prime numbers. Such a fraction is said to be in lowest terms.

*Example*

Express $\frac{56}{77}$ in lowest terms. We can use factoring or the Euclidean algorithm to find that the greatest common factor is 7. Thus,

$$\frac{56}{77} = \frac{7 \cdot 8}{7 \cdot 11} = \frac{8}{11},$$

and "$\frac{8}{11}$" is the fraction in lowest terms representing the given rational number.

Consider the two rational numbers $\frac{19}{27}$ and $\frac{200}{297}$. Are they equal? Since $19 \times 297 \neq 200 \times 27$, we conclude from the definition of equality that they are not equal. Since they are not equal, which rational is larger? One method to answer this question is as follows. First, express each rational number as a fraction with the same denominator. This is not difficult. Since $27 \times 297$ is a common multiple of 27 and 297, we can express each fraction with this common multiple as denominator.

$$\frac{19}{27} = \frac{19 \times 297}{27 \times 297} \quad \text{and} \quad \frac{200}{297} = \frac{27 \times 200}{27 \times 297},$$

or

$$\frac{19}{27} = \frac{5{,}643}{8{,}019} \quad \text{and} \quad \frac{200}{297} = \frac{5{,}400}{8{,}019}.$$

From our interpretation of rational numbers written in fractional notation, it is evident that the fraction with the larger numerator represents the larger rational. Thus, $\frac{19}{27}$ is *greater than* $\frac{200}{297}$, and this is expressed by $\frac{19}{27} > \frac{200}{297}$. Similarly, $\frac{200}{297}$ is *less than* $\frac{19}{27}$, and this is expressed by $\frac{200}{297} < \frac{19}{27}$.

In general, consider $\frac{m}{n}$ and $\frac{x}{y}$ where $m$, $n$, $x$, and $y$ are natural numbers. Since

$$\frac{m}{n} = \frac{my}{ny} \text{ and } \frac{x}{y} = \frac{nx}{ny},$$

we see that

$$\frac{m}{n} < \frac{x}{y} \text{ if and only if } my < nx.*$$

## Examples

1. $\frac{3}{17} < \frac{4}{21}$ since $3 \times 21 < 4 \times 17$; that is, $63 < 68$.
2. $\frac{5}{7} > \frac{42}{59}$ since $5 \times 59 > 7 \times 42$; that is, $295 > 294$.

In our attempt to find which of $\frac{19}{27}$ and $\frac{200}{297}$ was greater, we used the common multiple $27 \times 297$ of the denominators. This is not the least common multiple of the denominators. Using factor trees, we see that 27 is the G.C.F. of 27 and 297.

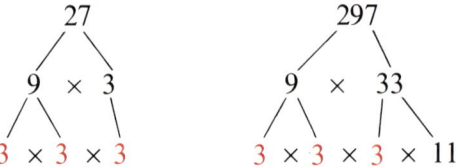

Hence, 297 is the L.C.M. of 27 and 297. We see that

$$\frac{19}{27} = \frac{11 \times 19}{11 \times 27} = \frac{209}{297};$$

thus, "$\frac{200}{297}$" and "$\frac{209}{297}$" are the fractional notations for the two rational numbers where 297 is the least common multiple of the denominators.

In general, for two rational numbers $\frac{m}{n}$ and $\frac{x}{y}$, we can express each in fractional notation where the denominator of each fraction is the the least common multiple of the denominators. The least common multiple of the denominators is called the least common denominator of the fractions and is denoted by L.C.D.

---

*This statement is true for the positive rationals; it will be necessary to reconsider some inequality properties when we introduce the negative rationals.

## 7.2 Equality and Inequality

### Example

For the fractions $\frac{7}{12}$ and $\frac{3}{44}$, the G.C.F. of the denominators is 4. Thus, $(44 \times 12) \div 4 = 132$ is the L.C.M. of 12 and 44; that is, 132 is the L.C.D. of the fractions. Since

$$\frac{7}{12} = \frac{11 \times 7}{11 \times 12} = \frac{77}{132} \quad \text{and} \quad \frac{3}{44} = \frac{3 \times 3}{3 \times 44} = \frac{9}{132}.$$

$\frac{77}{132}$ and $\frac{9}{132}$ are fractional notations for the two rationals using the L.C.D. for the denominator of each.

## Exercises I

For each pair of rational numbers in Exercises 1 through 6, determine if they are equal. If they are not equal, state which is greater.
1. (a) $\frac{11}{15}$ and $\frac{2}{3}$. (b) $\frac{4}{9}$ and $\frac{28}{63}$.
2. (a) $\frac{13}{18}$ and $\frac{65}{89}$. (b) $\frac{22}{35}$ and $\frac{131}{209}$.
3. (a) $\frac{19}{11}$ and $\frac{247}{143}$. (b) $\frac{63}{71}$ and $\frac{314}{355}$.
4. (a) $\frac{14}{9}$ and $\frac{306}{223}$. (b) $\frac{63}{114}$ and $\frac{575}{117}$.
5. (a) $\frac{12}{23}$ and $\frac{49}{91}$. (b) $\frac{3}{7}$ and $\frac{363}{847}$.
6. (a) $\frac{15}{7}$ and $\frac{301}{143}$. (b) $\frac{6}{11}$ and $\frac{119}{220}$.

For each set of rational numbers in Exercises 7 through 10, express each rational in the set in fractional notation using the L.C.D. of the given fractions.
7. (a) $\{\frac{7}{10}, \frac{9}{35}\}$. (b) $\{\frac{4}{13}, \frac{5}{11}\}$.
8. (a) $\{\frac{5}{46}, \frac{8}{54}\}$. (b) $\{\frac{3}{80}, \frac{23}{112}\}$.
9. (a) $\{\frac{23}{29}, \frac{6}{43}\}$. (b) $\{\frac{21}{34}, \frac{41}{68}\}$.
10. (a) $\{\frac{2}{3}, \frac{3}{5}, \frac{2}{7}\}$. (b) $\{\frac{5}{12}, \frac{5}{28}, \frac{15}{44}\}$.
11. Express each rational number as a fraction in lowest terms.
    (a) $\frac{10}{22}$. (b) $\frac{18}{12}$. (c) $\frac{212}{351}$. (d) $\frac{288}{451}$.
12. Express each rational number as a fraction in lowest terms.
    (a) $\frac{35}{75}$. (b) $\frac{48}{80}$. (c) $\frac{86}{215}$. (d) $\frac{184}{345}$.

## Exercises II

For each pair of rational numbers in Exercises 1 through 6, determine if they are equal. If they are not equal, state which is greater.

112                                                          *The Rational Numbers*

1. (a) $\frac{2}{3}$ and $\frac{16}{25}$. (b) $\frac{5}{2}$ and $\frac{91}{36}$.
2. (a) $\frac{8}{19}$ and $\frac{81}{180}$. (b) $\frac{19}{24}$ and $\frac{209}{264}$.
3. (a) $\frac{17}{12}$ and $\frac{81}{59}$. (b) $\frac{43}{87}$ and $\frac{129}{260}$.
4. (a) $\frac{5}{11}$ and $\frac{4}{187}$. (b) $\frac{6}{5}$ and $\frac{59}{49}$.
5. (a) $\frac{3}{7}$ and $\frac{68}{161}$. (b) $\frac{47}{61}$ and $\frac{329}{427}$.
6. (a) $\frac{62}{15}$ and $\frac{319}{76}$. (b) $\frac{43}{70}$ and $\frac{427}{697}$.

For each set of rational numbers in Exercises 7 through 10, express each rational in the set in fractional notation using the L.C.D. of the given fractions.

7. (a) $\{\frac{2}{9}, \frac{14}{15}\}$. (b) $\{\frac{3}{11}, \frac{2}{17}\}$.
8. (a) $\{\frac{11}{40}, \frac{13}{60}\}$. (b) $\{\frac{3}{16}, \frac{9}{40}\}$.
9. (a) $\{\frac{5}{14}, \frac{12}{35}\}$. (b) $\{\frac{8}{23}, \frac{6}{49}\}$.
10. (a) $\{\frac{16}{121}, \frac{40}{33}\}$. (b) $\{\frac{62}{137}, \frac{45}{61}\}$.
11. Discuss the following: Equals may be substituted for equals and the results are equal. Since the numerator of $\frac{6}{8}$ has 2 as a factor, we conclude by replacing $\frac{6}{8}$ by its equal $\frac{3}{4}$ that the numerator of $\frac{3}{4}$ has 2 as a factor.
12. Determine if the set of rational numbers is an ordered set.

## 7.3 Supplementary Exercises

In Exercises 1 through 5, determine the fractional representation in lowest terms for the number associated with the rational point on the number line or with the area of the shaded region.

1. (a)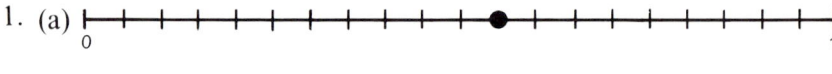

   (b)

2. (a)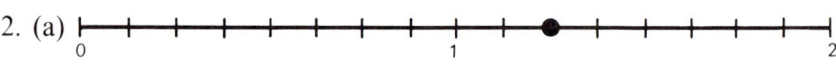

   (b)

3. (a)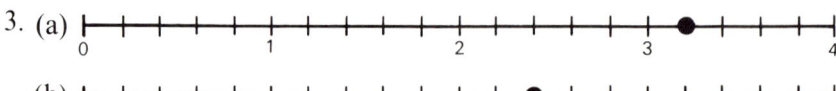

   (b)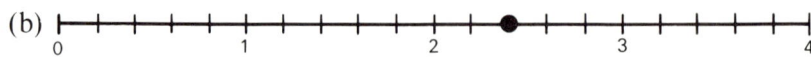

## 7.3 Supplementary Exercises

4. (a) 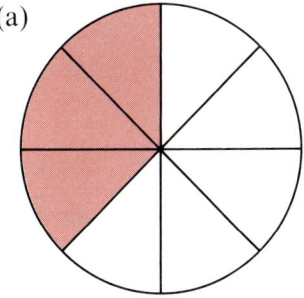  (b)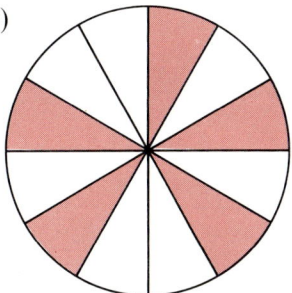

5. (a) 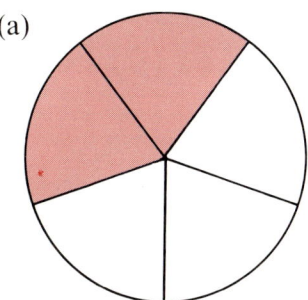  (b)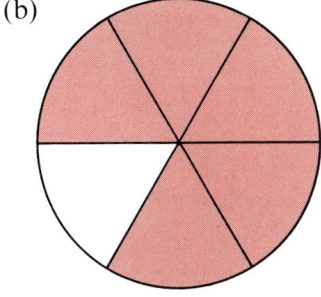

In Exercises 6 through 10, express each rational in the set in fractional notation using the L.C.D. of the given fractions.

6. (a) $\{\frac{5}{17}, \frac{4}{11}\}$. (b) $\{\frac{2}{3}, \frac{5}{6}, \frac{3}{8}\}$.
7. (a) $\{\frac{3}{5}, \frac{4}{9}\}$. (b) $\{\frac{1}{2}, \frac{2}{5}, \frac{3}{8}\}$.
8. (a) $\{\frac{6}{13}, \frac{1}{2}\}$. (b) $\{\frac{2}{3}, \frac{5}{9}, \frac{4}{7}\}$.
9. (a) $\{\frac{5}{8}, \frac{7}{12}\}$. (b) $\{\frac{2}{13}, \frac{1}{2}, \frac{4}{9}\}$.
10. (a) $\{\frac{3}{4}, \frac{11}{26}\}$. (b) $\{\frac{2}{3}, \frac{5}{6}, \frac{7}{18}\}$.

In each of the Exercises 11 through 15, determine if the pair of rationals are equal. If they are not, state which is greater.

11. (a) $\frac{11}{3}$ and $\frac{77}{21}$. (b) $\frac{13}{19}$ and $\frac{51}{76}$.
12. (a) $\frac{22}{13}$ and $\frac{199}{117}$. (b) $\frac{4}{9}$ and $\frac{3}{8}$.
13. (a) $\frac{15}{26}$ and $\frac{16}{27}$. (b) $\frac{53}{71}$ and $\frac{264}{355}$.
14. (a) $\frac{43}{59}$ and $\frac{473}{649}$. (b) $\frac{25}{39}$ and $\frac{74}{127}$.
15. (a) $\frac{481}{322}$ and $\frac{1443}{967}$. (b) $\frac{421}{537}$ and $\frac{1684}{2148}$.

# 8

$$3.17 + 5.42 = (3 + \tfrac{1}{10} + \tfrac{7}{100}) + (5 + \tfrac{4}{10} + \tfrac{2}{100})$$
$$= (3 + 5) + (\tfrac{1}{10} + \tfrac{4}{10}) + (\tfrac{7}{100} + \tfrac{2}{100})$$
$$= 8 + \tfrac{5}{10} + \tfrac{9}{100}$$
$$= 8.59$$

½

⅛

⅜

$$\begin{array}{r} 0.000 \\ 0.0000 \\ 0.00 \\ \hline n.nnnn \end{array}$$

$$\frac{c}{d} - \frac{a}{b} = \frac{bc}{bd} - \frac{ad}{bd}$$
$$= \frac{bc - ad}{bd}$$

¼

⅞

⅝

$$\frac{a}{b} + \frac{x}{y} = \frac{c}{d}$$

# ADDITION AND SUBTRACTION OF RATIONAL NUMBERS

## 8.1 Fractions

We not only use rational numbers for assigning measure to particular segments or regions but we also use this association to determine a proper definition for the sum of two rational numbers. If we have a segment which is $\frac{3}{17}$ of a unit in length and another which is $\frac{8}{17}$ of a unit in length, then the total length of the two segments would be $\frac{11}{17}$ of a unit. (See Figure 1.)

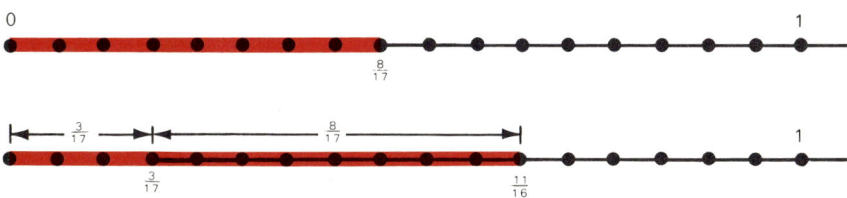

Figure 1. $\frac{3}{17} + \frac{8}{17} = \frac{11}{17}$.

For the two rationals $\frac{a}{b}$ and $\frac{c}{b}$, we *define* the sum $\frac{a}{b} + \frac{c}{b}$ to be the rational $\frac{a+c}{b}$. We should observe two things concerning this definition. First, as a consequence of the closure property of addition for the set of whole numbers, $a + c$ is a whole number; thus, the sum of two such rational numbers is a rational number. Second, this definition is consistent with the identification of $\frac{m}{1}$ with the whole number $m$. For example, to be consistent, since

$$\frac{3}{1} = 3 \quad \text{and} \quad \frac{5}{1} = 5,$$

we want the sum $\frac{3}{1} + \frac{5}{1}$ to be 8. From the definition of sum for two rationals, we have

$$\frac{3}{1} + \frac{5}{1} = \frac{8}{1}.$$

Since $\frac{8}{1} = 8$, we obtain the desired sum with the given definition of addition. (A common error is to add denominators; we can easily show that this would be incompatible with the identification $\frac{m}{1} = m$ and the definition of addition for whole numbers.)

For *any* two rational numbers $\frac{a}{b}$ and $\frac{c}{d}$,

$$\frac{a}{b} = \frac{ad}{bd} \quad \text{and} \quad \frac{c}{d} = \frac{bc}{bd}.$$

Using the definition of addition,

$$\frac{a}{b} + \frac{c}{d} = \frac{ad}{bd} + \frac{bc}{bd}$$

$$= \frac{ad + bc}{bd}.$$

From the closure properties for addition and multiplication of whole numbers, we see that the sum of any two rational numbers is a rational number; thus, *the set of rational numbers is closed with respect to addition.*

## Examples

1. $\dfrac{5}{7} + \dfrac{7}{8} = \dfrac{(5 \cdot 8) + (7 \cdot 7)}{7 \cdot 8} = \dfrac{40 + 49}{56} = \dfrac{89}{56}.$

2. $\dfrac{3}{4} + \dfrac{5}{12} = \dfrac{(3 \cdot 12) + (4 \cdot 5)}{4 \cdot 12} = \dfrac{36 + 20}{48} = \dfrac{56}{48}.$

3. $\dfrac{2}{5} + \dfrac{3}{8} = \dfrac{(2 \cdot 8) + (5 \cdot 3)}{5 \cdot 8} = \dfrac{16 + 15}{40} = \dfrac{31}{40}.$

This procedure for adding numbers using fractional notation is often simplified if we express the rational numbers in fractional notation using the least common denominator of the fractions. Consider carefully the following examples.

## 8.1 Fractions

**Examples**

1. $\dfrac{3}{4} + \dfrac{5}{12} = \dfrac{9}{12} + \dfrac{5}{12} = \dfrac{14}{12}.$  See Example 2 above and observe that $\dfrac{56}{48} = \dfrac{14}{12}.$

2. $\dfrac{3}{10} + \dfrac{7}{15} = \dfrac{3}{30} + \dfrac{14}{30} = \dfrac{17}{30}.$

3. $\dfrac{3}{7} + \dfrac{2}{9} = \dfrac{27}{63} + \dfrac{14}{63} = \dfrac{41}{63}.$

4. $\dfrac{11}{24} + \dfrac{5}{36} = \dfrac{33}{72} + \dfrac{10}{72} = \dfrac{43}{72}.$

For arithmetical purposes as well as for purposes of application, it is generally desirable to express the sum of rationals as a fraction in lowest terms. This task is usually made less difficult if the L.C.D. is used.

**Examples**

1. $\dfrac{3}{4} + \dfrac{5}{12} = \dfrac{9}{12} + \dfrac{5}{12} = \dfrac{14}{12} = \dfrac{2 \times 7}{2 \times 6} = \dfrac{7}{6}.$

2. $\dfrac{4}{21} + \dfrac{9}{14} = \dfrac{8}{42} + \dfrac{27}{42} = \dfrac{35}{42} = \dfrac{7 \times 5}{7 \times 6} = \dfrac{5}{6}.$

It is not difficult to show that the set of rational numbers has the *commutative property of addition.* Let us prove that $\frac{a}{b} + \frac{c}{d} = \frac{c}{d} + \frac{a}{b}.$

**Step 1.** $\quad \dfrac{a}{b} + \dfrac{c}{d} = \dfrac{ad + bc}{bd} \quad$ Definition of addition.

**Step 2.** $\quad \dfrac{c}{d} + \dfrac{a}{b} = \dfrac{cb + da}{db} \quad$ Definition of addition.

$\qquad\qquad\qquad = \dfrac{da + cb}{db} \quad$ Commutative property of addition for whole numbers.

$\qquad\qquad\qquad = \dfrac{ad + bc}{bd} \quad$ Commutative property of multiplication for whole numbers.

**Step 3.** From Steps 1 and 2, we conclude that

$$\frac{a}{b} + \frac{c}{d} = \frac{c}{d} + \frac{a}{b}.$$

To prove that the set of rational numbers also has the *associative property of addition* is not too difficult, but it is rather tedious. Several numerical examples should serve to convince us that the set does have the associative property of addition.

For any two rational numbers $\frac{a}{b}$ and $\frac{c}{d}$, if there exists a unique rational $\frac{x}{y}$ such that $\frac{a}{b} + \frac{x}{y} = \frac{c}{d}$, then from the definition of subtraction $\frac{x}{y}$ is the *difference* $\frac{c}{d} - \frac{a}{b}$. If $\frac{c}{d}$ is the length of a segment consisting of two segments, one with $\frac{a}{b}$ as length and one with $\frac{x}{y}$ as length, then it is geometrically obvious that (until we extend the number system to include negative numbers) the number $\frac{c}{d}$ must be greater than $\frac{a}{b}$ for the indicated difference to exist. In other words, as with the set of whole numbers, we do not (yet) have closure with respect to subtraction for the set of rational numbers.

To find the difference $\frac{15}{17} - \frac{3}{17}$, we need to find the rational $\frac{x}{y}$, if it exists, such that $\frac{3}{17} + \frac{x}{y} = \frac{15}{17}$. From our definition of addition, it is immediately obvious that if $y = 17$ and $x = 12$ then $\frac{x}{y} = \frac{12}{17}$ is the difference. Geometrically, if we have a segment $\frac{15}{17}$ of a unit in length and if we "take away" a segment $\frac{3}{17}$ of a unit in length, then we have a segment $\frac{12}{17}$ of a unit in length. (See Figure 2.)

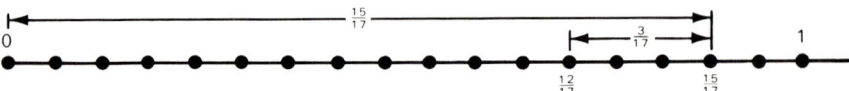

Figure 2. $\frac{15}{17} - \frac{3}{17} = \frac{12}{17}$.

If $\frac{c}{b} > \frac{a}{b}$ then $\frac{c}{b} - \frac{a}{b}$ is the rational $\frac{c-a}{b}$. In general, for rationals $\frac{a}{b}$ and $\frac{c}{d}$ where $\frac{c}{d} > \frac{a}{b}$, we obtain a common denominator and subtract numerators. Thus,

$$\frac{c}{d} - \frac{a}{b} = \frac{bc}{bd} - \frac{ad}{bd}$$

$$= \frac{bc - ad}{bd}.$$

Recall, if $\frac{c}{d} > \frac{a}{b}$ then $bc > ad$ and $bc - ad$ is a natural number.

As with addition, it is advantageous, as well as standard practice,

# 8.1 Fractions

to use the L.C.D. of the fractions when performing the subtraction operation.

**Examples**

1. $\dfrac{2}{3} - \dfrac{3}{7} = \dfrac{14-9}{21} = \dfrac{5}{21}.$

2. $\dfrac{9}{14} - \dfrac{4}{21} = \dfrac{27}{42} - \dfrac{8}{42} = \dfrac{19}{42}.$

3. $\dfrac{8}{9} - \dfrac{5}{6} = \dfrac{16}{18} - \dfrac{15}{18} = \dfrac{1}{18}.$

4. $\dfrac{11}{12} - \dfrac{7}{18} = \dfrac{33}{36} - \dfrac{14}{36} = \dfrac{19}{36}.$

Although we can pair any rational number with a point on the number line, we shall see in Chapter 12 that there are points on the number line which are not associated with any rational number. In other words, we shall find that not all points on the number line are rational points and not all segments have rational length. Fortunately, we can prove that it is possible to measure the length of such a segment to any desired degree of accuracy with a rational number.

## Exercises I

In Exercises 1 through 6, find the indicated sums and express your answer as a fraction in lowest terms.

1. (a) $\dfrac{2}{5} + \dfrac{3}{11}.$ (b) $\dfrac{4}{7} + \dfrac{5}{8}.$
2. (a) $\dfrac{3}{16} + \dfrac{5}{12}.$ (b) $\dfrac{5}{7} + \dfrac{10}{21}.$
3. (a) $\dfrac{6}{11} + \dfrac{3}{22}.$ (b) $\dfrac{7}{15} + \dfrac{9}{20}.$
4. (a) $\dfrac{17}{20} + \dfrac{19}{30}.$ (b) $\dfrac{6}{35} + \dfrac{8}{63}.$
5. (a) $\left(\dfrac{2}{9} + \dfrac{5}{12}\right) + \dfrac{7}{20}.$ (b) $\dfrac{3}{10} + \left(\dfrac{7}{15} + \dfrac{10}{21}\right).$
6. (a) $\left(\dfrac{11}{14} + \dfrac{5}{6}\right) + \dfrac{4}{15}.$ (b) $\dfrac{11}{14} + \left(\dfrac{5}{6} + \dfrac{4}{15}\right).$

In Exercises 7 and 8, find the indicated differences and express your answer as a fraction in lowest terms.

7. (a) $\dfrac{17}{15} - \dfrac{8}{9}.$ (b) $\dfrac{23}{12} - \dfrac{11}{14}.$
8. (a) $\dfrac{5}{9} - \dfrac{2}{15}.$ (b) $\dfrac{41}{20} - \dfrac{14}{15}.$

In Exercises 9 and 10, perform the indicated operations and express your answer as a fraction in lowest terms.

9. $(\frac{5}{12} + \frac{23}{24}) - (\frac{2}{5} + \frac{3}{4}) + (\frac{5}{6} + \frac{2}{3})$.
10. $(\frac{12}{5} - \frac{3}{4}) + (\frac{3}{10} + \frac{5}{3}) - (\frac{1}{6} + \frac{4}{15})$.

### Exercises II

1. Use the definition that $\frac{0}{d} = 0$ for any natural number $d$ and the definition of addition for rational numbers to prove that

$$\frac{a}{b} + 0 = \frac{a}{b}.$$

2. A common error made when adding $\frac{a}{b}$ and $\frac{c}{d}$ is to add the numerators and denominators and obtain $\frac{a+c}{b+d}$. Show by using the rationals $\frac{5}{1}$ and $\frac{3}{1}$ that this would be inconsistent with our previous knowledge for whole numbers and the definition of fractions.

In Exercises 3 through 9, perform the indicated operations and express your answer as a fraction in lowest terms.

3. $(\frac{2}{3} + \frac{5}{7}) + (\frac{5}{9} - \frac{2}{7})$.
4. $[(\frac{3}{5} + \frac{7}{9}) - (\frac{1}{3} + \frac{2}{9})] + (\frac{7}{15} - \frac{1}{5})$.
5. $(\frac{3}{4} - \frac{2}{11}) + (\frac{3}{22} + \frac{1}{2}) + (\frac{5}{11} - \frac{1}{4})$.
6. $(\frac{2}{3} + \frac{5}{6}) + [(\frac{1}{2} + \frac{3}{7}) - (\frac{2}{3} + \frac{1}{6})]$.
7. $(\frac{3}{2} - \frac{5}{12}) + (\frac{5}{6} - \frac{1}{3}) + (\frac{3}{4} + \frac{7}{12})$.
8. $[(\frac{7}{24} + \frac{1}{2}) - (\frac{2}{3} - \frac{1}{4})] + (\frac{1}{6} + \frac{5}{12})$.
9. $(\frac{4}{9} + \frac{2}{27}) + (\frac{4}{3} - \frac{5}{6}) + (\frac{7}{18} + \frac{5}{54})$.
10. For any three rational numbers $\frac{a}{b}$, $\frac{c}{d}$, and $\frac{e}{f}$, prove that

$$\left(\frac{a}{b} + \frac{c}{d}\right) + \frac{e}{f} = \frac{a}{b} + \left(\frac{c}{d} + \frac{e}{f}\right).$$

## 8.2 Mixed Numerals

From our previous definitions, it follows that

$$\frac{18}{13} = \frac{13}{13} + \frac{5}{13} = 1 + \frac{5}{13}.$$

## 8.2 Mixed Numerals

With improper fractions such as "$\frac{18}{13}$," it is often more convenient, as well as important in certain applications, to express the rational as the sum of a whole number and rational number written as a proper fraction. For example, we can quickly conclude that $5 + \frac{17}{19}$ is less than $6 + \frac{1}{17}$, but in the equivalent fractional notation it may not be so obvious that $\frac{112}{19} < \frac{103}{17}$.

To represent the sum $1 + \frac{5}{13}$, we use the notation "$1\frac{5}{13}$"; this is called the mixed numeral notation for the rational number $\frac{18}{13}$. To find the mixed numeral notation for, say, $\frac{87}{13}$, we first need to find the greatest whole number less than $\frac{87}{13}$. Since $6 \times 13 = 78$ is the greatest multiple of 13 less than 87, 6 is the greatest whole number less than $\frac{87}{13}$. Thus, we see that

$$\frac{87}{13} = \frac{78}{13} + \frac{11}{13} = 6 + \frac{11}{13} = 6\frac{11}{13}.$$

In general, if "$\frac{a}{b}$" is an improper fraction, then $\frac{a}{b} = q + \frac{r}{b}$ where $q$ is the incomplete quotient and $r$ is the remainder in the long division of $a$ by $b$. Since $r < b$, the fraction $\frac{r}{b}$ is a proper fraction provided $r \neq 0$; if $r = 0$, then $b$ is a factor of $a$, and $\frac{a}{b}$ is an improper fraction representing the whole number $q$.

### Examples

1. $\dfrac{41}{12} = 3 + \dfrac{5}{12} = 3\dfrac{5}{12}.$

2. $\dfrac{86}{14} = 6 + \dfrac{2}{14} = 6 + \dfrac{1}{7} = 6\dfrac{1}{7}.$

3. $\dfrac{96}{16} = 6 + \dfrac{0}{16} = 6.$

Since $2\frac{9}{5} = 2 + \frac{9}{5} = \frac{10}{5} + \frac{9}{5} = \frac{19}{5}$, we see that "$2\frac{9}{5}$" denotes the rational number $\frac{19}{5}$; however, we do not call "$2\frac{9}{5}$" the mixed numeral notation for $\frac{19}{5}$. We only use "mixed numeral notation" to refer to "$q\frac{r}{b}$" where $\frac{r}{b}$ is a proper fraction in lowest terms. Thus, "$3\frac{4}{5}$" is the mixed numeral notation for $\frac{19}{5}$. Requiring "$\frac{r}{b}$" to be a proper fraction in the mixed numeral notation is not universal practice, but we do this for three reasons. (1) We may then talk about *the* mixed numeral notation for a rational number such as $\frac{19}{5}$. (2) We may discuss the method for expressing an improper fraction in mixed numeral notation. (3) Most important, the definition indicates the appropriate use of the notation.

To find the sum (or difference) of $8\frac{3}{7}$ and $4\frac{1}{3}$, we could express each rational number as an improper fraction and proceed by the methods discussed in the last section. However, it is usually easier to obtain the sum as follows.

$$8\frac{3}{7} + 4\frac{1}{3} = \left(8 + \frac{3}{7}\right) + \left(4 + \frac{1}{3}\right)$$

$$= (8 + 4) + \left(\frac{3}{7} + \frac{1}{3}\right)$$

$$= 12 + \left(\frac{9}{21} + \frac{7}{21}\right)$$

$$= 12 + \frac{16}{21}$$

$$= 12\frac{16}{21}.$$

This technique is simplified considerably by using "vertical" addition.

$$\text{Add:} \quad 8\frac{3}{7} = 8\frac{9}{21}$$

$$4\frac{1}{3} = 4\frac{7}{21}$$

$$\text{Sum} \quad \overline{12\frac{16}{21}}.$$

The following examples should indicate the variations of this important technique for adding rational numbers written in mixed numeral notation.

*Examples*

1. Add: $\quad 7\frac{2}{5} = 7\frac{8}{20} \qquad$ 2. Subtract: $\quad 17\frac{3}{5} = 17\frac{12}{20}$

$\qquad\qquad\qquad 3\frac{1}{4} = 3\frac{5}{20} \qquad\qquad\qquad\qquad\quad 6\frac{1}{4} = 6\frac{5}{20}$

$\qquad\quad \text{Sum} \quad\;\; \overline{10\frac{13}{20}} \qquad\qquad \text{Difference} \qquad\;\; \overline{11\frac{7}{20}}$

## 8.2 Mixed Numerals

3. Add: $9\dfrac{2}{3} = 9\dfrac{4}{6}$

$\phantom{\text{Add: }}8\dfrac{1}{2} = 8\dfrac{3}{6}$

Sum $\phantom{\text{Add: }}17\dfrac{7}{6} = 17 + 1 + \dfrac{1}{6} = 18\dfrac{1}{6}$

4. Subtract: $14\dfrac{2}{5} = 14\dfrac{4}{10} = 13\dfrac{14}{10}$

$\phantom{\text{Subtract: }}10\dfrac{1}{2} = 10\dfrac{5}{10} = 10\dfrac{5}{10}$

Difference $\phantom{\text{Subtract: aaa}}3\dfrac{9}{10}$

In Example 4, it is necessary to change the notation in the minuend so that the fraction represents a rational number greater than the rational represented by the fraction in the subtrahend.

## Exercises I

1. Express each rational number as an improper fraction.
   (a) $3\dfrac{5}{7}$. (b) $4\dfrac{8}{9}$. (c) $6\dfrac{5}{11}$. (d) $7\dfrac{5}{8}$.
2. Express each rational number in mixed numeral notation.
   (a) $\dfrac{38}{9}$. (b) $\dfrac{29}{5}$. (c) $\dfrac{49}{12}$. (d) $\dfrac{54}{17}$.

In Exercises 3 through 6, find the indicated sums and express your answer in mixed numeral notation.

3. (a) $6\dfrac{2}{3}$    (b) $7\dfrac{2}{5}$    (c) $8\dfrac{5}{7}$
   $\phantom{(a) }4\dfrac{1}{2}$      $8\dfrac{3}{10}$     $4\dfrac{2}{3}$

4. (a) $4\dfrac{2}{9}$    (b) $11\dfrac{3}{7}$    (c) $8\dfrac{5}{12}$
   $\phantom{(a) }8\dfrac{2}{3}$      $18\dfrac{5}{9}$     $6\dfrac{4}{15}$

5. (a) $8\dfrac{11}{16}$    (b) $17\dfrac{5}{11}$    (c) $14\dfrac{7}{15}$
   $\phantom{(a) }3\dfrac{1}{4}$      $6\dfrac{3}{22}$     $11\dfrac{3}{20}$
   $\phantom{(a) }2\dfrac{1}{2}$      $5\dfrac{1}{3}$     $23\dfrac{2}{3}$

6. (a) $3\dfrac{2}{3}$    (b) $4\dfrac{3}{5}$    (c) $9\dfrac{5}{8}$
   $\phantom{(a) }8\dfrac{5}{8}$      $5\dfrac{1}{4}$     $5\dfrac{1}{3}$
   $\phantom{(a) }2\dfrac{1}{6}$      $8\dfrac{2}{3}$     $4\dfrac{1}{2}$

In Exercises 7 through 10, find the indicated differences and express your answer in mixed numeral notation.

7. (a) $11\frac{2}{3}$    (b) $4\frac{3}{7}$    (c) $15\frac{4}{5}$
       $\underline{5\frac{1}{2}}$        $\underline{1\frac{1}{6}}$        $\underline{9\frac{3}{10}}$

8. (a) $8\frac{1}{3}$    (b) $11\frac{2}{7}$    (c) $12\frac{3}{7}$
       $\underline{4\frac{2}{5}}$        $\underline{6\frac{3}{4}}$        $\underline{8\frac{1}{2}}$

9. (a) $5\frac{8}{9}$    (b) $6\frac{2}{3}$    (c) $10\frac{2}{5}$
       $\underline{2\frac{3}{5}}$        $\underline{4\frac{1}{8}}$        $\underline{4\frac{1}{9}}$

10. (a) $2\frac{1}{7}$    (b) $11\frac{2}{3}$    (c) $7\frac{1}{8}$
        $\underline{1\frac{4}{9}}$        $\underline{5\frac{2}{15}}$        $\underline{3\frac{5}{6}}$

## Exercises II

1. Express each rational number as an improper fraction.
   (a) $4\frac{1}{5}$. (b) $2\frac{8}{9}$. (c) $6\frac{7}{15}$. (d) $4\frac{2}{13}$.
2. Express each rational number as an improper fraction.
   (a) $11\frac{1}{3}$. (b) $12\frac{4}{17}$. (c) $9\frac{5}{16}$. (d) $11\frac{2}{13}$.
3. Express each rational number in mixed numeral notation.
   (a) $\frac{231}{17}$. (b) $\frac{42}{5}$. (c) $\frac{81}{17}$. (d) $\frac{86}{23}$.
4. Express each rational number in mixed numeral notation.
   (a) $\frac{421}{38}$. (b) $\frac{396}{42}$. (c) $\frac{41}{8}$. (d) $\frac{322}{16}$.

In Exercises 5 through 10 perform the indicated operations and express your answer in mixed numeral notation.

5. $5\frac{2}{3} + 8\frac{5}{6} + 7\frac{1}{5}$.        6. $11\frac{1}{8} + 3\frac{2}{7} + 5\frac{1}{4}$.
7. $16\frac{2}{7} + 8\frac{1}{3} + 9\frac{5}{6}$.        8. $(12\frac{3}{8} + 2\frac{1}{5}) - 4\frac{2}{3}$.
9. $(17\frac{2}{3} - 8\frac{1}{7}) + 3\frac{3}{8}$.        10. $(16\frac{1}{8} - 4\frac{2}{5}) - 2\frac{3}{4}$.

## 8.3 Decimals

It can be quite an arithmetic task to find the sum of two rational numbers written in fractional notation or in mixed numeral notation. For example, to express the sum

$$\frac{217{,}421}{1{,}698{,}434} + \frac{23{,}123}{87{,}425}$$

## 8.3 Decimals

as a fraction in lowest terms would require rather extensive calculations. Fortunately, most calculations of this type can be avoided by using the *decimal notation* for rational numbers; this important notation was developed in the sixteenth century by Simon Stevin of Belgium.

The decimal notation is an extension of the positional notation for the counting numbers. We use "8.237" to represent the sum

$$8 + \frac{2}{10} + \frac{3}{100} + \frac{7}{1,000}.$$

The "period" between "8" and "2" is the *separatrix* in the notation and is called the decimal point.* The first position to the right of the decimal point (in base ten) is called the *tenths' position*, the second is called the *hundreths' position*, the third is called the *thousandths' position*, etc. (See Figure 3.)

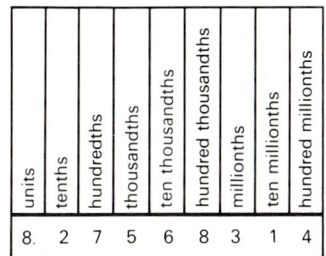

Figure 3. *Decimals.*

Our work with decimals is simplified if we introduce some elementary ideas concerning *exponents*. If $n$ is any natural number greater than 1, we define $a^n$ to be the product of $n$ numbers each of which is $a$; that is,

$$a^n = \underbrace{a \times a \times a \times a \times \cdots \times a}_{n \text{ terms}}.$$

Furthermore, we define $a^1$ to be $a$. For example, $2^3 = 2 \times 2 \times 2$, $10^2 = 10 \times 10$, and $10^4 = 10 \times 10 \times 10 \times 10$. We call $a^n$ the *n*th power of $a$.

Since $10^2 = 10 \times 10$ and $10^3 = 10 \times 10 \times 10$,

---

*The use of a decimal point for the separatrix is not universal. For example, in some countries a "raised" period is used and in others a comma is used.

$$10^2 \times 10^3 = (10 \times 10) \times (10 \times 10 \times 10)$$
$$= 10 \times 10 \times 10 \times 10 \times 10$$
$$= 10^5.$$

In general, if $m$ and $n$ are natural numbers, then

$$a^m \times a^n = a^{m+n}.$$

The decimal "4.23" (read, "four and twenty-three hundredths," or "four-point-two-three") represents

$$4 + \frac{2}{10} + \frac{3}{100}.$$

Since

$$4 + \frac{2}{10} + \frac{3}{100} = 4 + \frac{20}{100} + \frac{3}{100} = 4\frac{23}{100},$$

"$4\frac{23}{100}$" is the mixed numeral notation for the rational number whose decimal representation is "4.23". The improper fraction "$\frac{423}{100}$" also denotes the same rational number.

The decimal 4.23 is called a *two-place* decimal; similarly, 8.237 is called a *three-place* decimal. The term *n*-place decimal indicates that $n$ positions to the right of the decimal are used in denoting the rational number. It should be realized that the two-place decimal 4.23 and the four-place decimal 4.2300 represent the same rational number. (In physical applications, it is useful to take advantage of these different notations and make a distinction by introducing the concept of "significant figures"; however, in arithmetic no such distinction is made.)

From the definition of the decimal notation, an *n*-place decimal represents a finite (or terminating) sum. For this reason, an *n*-place decimal is often called a finite decimal, or terminating decimal. In Chapter 12, we shall make an extension of the decimal notation and introduce infinite decimals, or nonterminating decimals, but at present we shall consider all decimals to be *n*-place (finite) decimals.

Any rational number denoted by a decimal can also be denoted by a fraction whose denominator is a power of 10; we call such fractions decimal fractions. It is also true that any rational number written as a decimal fraction can be expressed in decimal notation. We shall find in the next section that not every rational number can be written as a (simple finite) decimal but, fortunately, any rational such as $\frac{217,421}{1,698,434}$ can be approximated to any required

## 8.3 Decimals

degree of accuracy by a finite decimal. This fact makes the decimal notation for rational numbers quite satisfactory in almost any application using rational numbers.

### Examples

1. 3.271 (read "three and two hundred seventy-one thousandths" or "three-point-two-seven-one") can be written in the mixed numeral notation "$3\frac{271}{1,000}$" or as the decimal fraction "$\frac{3,271}{1,000}$." Notice that "3.271" is a three-place decimal and the denominator of the decimal fraction is $10^3$.
2. 5.0043 (read "five and forty-three ten thousandths" or "five-point-zero-zero-four-three") can be written in the mixed numeral notation "$5\frac{43}{10,000}$" or as the decimal fraction "$\frac{50,043}{10,000}$."
3. 6.8 (read "six and eight-tenths" or "six-point-eight") can be written as $6\frac{8}{10}$ which is $\frac{68}{10}$ as a decimal fraction.
4. $\frac{671}{100} = 6\frac{71}{100} = 6.71$.
5. $\frac{6,429}{100} = 64\frac{29}{100} = 64.29$.

Finding the sum $3.42 + 5.27$ is equivalent to finding the sum $(3 + \frac{4}{10} + \frac{2}{100}) + (5 + \frac{2}{10} + \frac{7}{100})$. Thus,

$$3.42 + 5.27 = \left(3 + \frac{4}{10} + \frac{2}{100}\right) + \left(5 + \frac{2}{10} + \frac{7}{100}\right)$$

$$= (3 + 5) + \left(\frac{4}{10} + \frac{2}{10}\right) + \left(\frac{2}{100} + \frac{7}{100}\right)$$

$$= 8 + \frac{6}{10} + \frac{9}{100}$$

$$= 8.69.$$

It should be obvious that an easier method to obtain the above sum is to write the decimals in a column with the decimal points in a vertical line and then add each column.

$$\begin{array}{r} 3.42 \\ 5.27 \\ \hline 8.69 \end{array}$$

To determine the sum $2.738 + 2.0451 + 3.5 + 4.15$ and express it in decimal notation, we can "fill in" zeros as shown below to facilitate column addition. However, this technique is usually avoided after acquiring some familiarity with the decimal notation.

$$\begin{array}{r} 1\ 11\phantom{0} \\ 2.7380 \\ 2.0451 \\ 3.5000 \\ 4.1500 \\ \hline 12.4331 \end{array}$$

In the preceding illustration, for example, it should be noted that since

$$\frac{8}{1{,}000} + \frac{5}{1{,}000} = \frac{13}{1{,}000}$$

$$= \frac{10}{1{,}000} + \frac{3}{1{,}000}$$

$$= \frac{1}{100} + \frac{3}{1{,}000},$$

we write "3" in the thousandths' place and "carry 1" to be added in the hundredths' column.

To determine the difference 34.462 − 22.81 and express it in decimal notation, we write the decimals in a column and proceed as we did to find the difference of two given natural numbers.

$$\begin{array}{ll} \text{Subtract} & 34.462 \\ & \underline{22.81\phantom{0}} \\ \text{Difference} & 11.652 \end{array}$$

## Exercises I

In Exercises 1 through 4, express each rational number in mixed numeral notation.

1. (a) 2.5.     (b) 4.25.     (c) 2.125.
2. (a) 3.375.     (b) 6.4.     (c) 8.2.
3. (a) 5.675.     (b) 7.6.     (c) 10.12.
4. (a) 8.3.     (b) 6.1.     (c) 4.875.

In Exercises 5 through 8, determine the sum of the given rationals.

5. (a) 34.68
        21.53

   (b) 26.11
       81.42

   (c) 3.921
       4.278

## 8.4 Approximating Rations with Decimals

6. (a) 64.125
      23.481
      ———
   (b) 8.425
       7.938
       ———
   (c) 7.612
       3.869
       ———

7. (a) 2.413
      5.718
      3.216
      ———
   (b) 4.007
       8.306
       2.514
       ———
   (c) 8.213
       4.728
       8.139
       ———

8. (a) 23.46
      17.81
      51.32
      21.69
      ———
   (b) 7.281
       4.310
       6.205
       8.898
       ———
   (c) 8.261
       3.821
       2.729
       4.655
       ———

In Exercises 9 and 10, determine the difference of the given rationals.

9. (a) 68.243
      23.141
      ———
   (b) 47.238
       11.883
       ———
   (c) 8.2415
       3.2877
       ———

10. (a) 23.217
       11.555
       ———
    (b) 17.814
        12.233
        ———
    (c) 9.4106
        3.2871
        ———

### Exercises II

Perform the indicated operations in each of the following exercises.
1. $23.21 + 4.68 + 5.11 + 8.31$.
2. $6.81 + 5.2 + 1.236 + 5.01$.
3. $8.1 + 5.202 + 3.2146$.
4. $6.15 + 31.4 + 53.712$.
5. $23.001 + 2.25 + 4.6 + 3.1$.
6. $25.712 - 11.281$.
7. $81.612 - 55.21$.
8. $143.81 - 71.639$.
9. $46.01 - 39.002$.
10. $57.142 - 33.2847$.

## 8.4 Approximating Rationals with Decimals

From our previous examples and exercises, we found that every $n$-place decimal represents a rational number. Unfortunately, it is not true that every rational number can be expressed as an $n$-place decimal. For example, if $\frac{1}{3}$ could be expressed as an $n$-place decimal then $\frac{1}{3} = \frac{q}{10^n}$ for some natural numbers $q$ and $n$. By the definition of equality for rationals, $\frac{1}{3} = \frac{q}{10^n}$ if and only if

$$3q = 10^n.$$

The equality would imply that 3 is a factor of $10^n$; however, this is not true. Hence, $\frac{1}{3}$ cannot be expressed as an $n$-place decimal.

Although $\frac{1}{3}$ is not equal to, say, $\frac{q}{1,000}$ for any natural number $q$, we can find the greatest natural number $q$ such that

$$\frac{q}{1,000} \leq \frac{1}{3},$$

or such that

$$3q \leq 1,000.$$

This is equivalent to finding the greatest multiple, $3q$, of 3 less than or equal to 1,000. The number $q$, of course, can be found by the long division algorithm.

$$\begin{array}{r} 3\,3\,3 = q \\ 3\,\overline{\smash{\big)}\,1{,}0\,0\,0} \\ \underline{9}\phantom{00} \\ 1\,0\phantom{0} \\ \underline{9}\phantom{0} \\ 1\,0 \\ \underline{9} \\ 1 = \text{remainder} \end{array}$$

Thus, $\frac{333}{3,000}$ is the greatest rational of the form $\frac{q}{1,000}$ less than or equal to $\frac{1}{3}$. Similarly, .333 (or 0.333 as it is more usually written) represents the best three-place decimal approximation of $\frac{1}{3}$ which is less than $\frac{1}{3}$. Since

$$\frac{1}{3} - \frac{333}{1,000} = \frac{1,000}{3,000} - \frac{999}{3,000}$$

$$= \frac{1}{3,000},$$

the three-place decimal 0.333 represents a rational approximation of $\frac{1}{3}$ which differs from $\frac{1}{3}$ by only $\frac{1}{3,000}$.

We can prove that if the $(n + 1)$-place decimal approximation

## 8.4 Approximating Rationals with Decimals

obtained above is not equal to the *n*-place decimal approximation, then it must be a better approximation of the given rational. In fact, the difference between the rational and its decimal approximation can be made arbitrarily small. (See Chapter 12.)

### Examples

1. To find the best four-place decimal approximation of $\frac{1}{3}$ which is less than or equal to $\frac{1}{3}$, we seek the greatest natural number $q$ such that

$$\frac{q}{10,000} \leq \frac{1}{3} \quad \text{or} \quad 3q \leq 10,000.$$

By long division, we obtain $q = 3,333$ and $r = 1$. Thus, $\frac{3,333}{10,000} = 0.3333$ is the best four-place decimal approximation of $\frac{1}{3}$ which is less than or equal to $\frac{1}{3}$.

2. To find the best four-place decimal approximation of $\frac{23}{127}$ which is less than or equal to that number, we find the greatest natural number $q$ such that

$$\frac{q}{10,000} \leq \frac{23}{127} \quad \text{or} \quad 127q \leq 230,000.$$

Since

$$\begin{array}{r} 1{,}8\,1\,1 = q \\ 127\overline{\smash{)}2\,3\,0{,}0\,0\,0} \\ \underline{1\,2\,7\phantom{{,}0\,0\,0}} \\ 1\,0\,3\,0\phantom{\,0\,0} \\ \underline{1\,0\,1\,6\phantom{\,0\,0}} \\ 1\,4\,0\phantom{\,0} \\ \underline{1\,2\,7\phantom{\,0}} \\ 1\,3\,0 \\ \underline{1\,2\,7} \\ 3 = \text{remainder} \end{array}$$

it follows that $\frac{1,811}{10,000} = 0.1811$ is the best four-place decimal approximation of $\frac{23}{127}$ which is less than or equal to the given rational.

3. Find the best four-place decimal approximation of $\frac{11}{17}$ which is less than or equal to the rational number.

$$\begin{array}{r}
6{,}470 = q\phantom{000}\\
17\overline{\smash{\big)}110{,}000}\\
\underline{102}\phantom{,000}\\
80\phantom{00}\\
\underline{68}\phantom{00}\\
120\phantom{0}\\
\underline{119}\phantom{0}\\
10 = \text{remainder}
\end{array}$$

Thus, $\frac{6{,}470}{10{,}000} = 0.6470$ is the best four-place decimal approximation of $\frac{11}{17}$ which is less than or equal to the given rational number.

4. Find the best four-place decimal approximation of $\frac{197}{800}$ which is less than or equal to the rational number.

$$\begin{array}{r}
2{,}462 = q\phantom{000}\\
800\overline{\smash{\big)}1{,}970{,}000}\\
\underline{1600}\phantom{,000}\\
3700\phantom{00}\\
\underline{3200}\phantom{00}\\
5000\phantom{0}\\
\underline{4800}\phantom{0}\\
2000\\
\underline{1600}\\
400 = \text{remainder}
\end{array}$$

Answer: $\dfrac{2{,}462}{10{,}000} = 0.2462.$

5. Find the best four-place decimal approximation of $\frac{18}{7}$ which is less than or equal to the rational number.

$$\begin{array}{r}
25{,}714 = q\phantom{000}\\
7\overline{\smash{\big)}180{,}000}\\
\underline{14}\phantom{0{,}000}\\
40\phantom{000}\\
\underline{35}\phantom{000}\\
50\phantom{00}\\
\underline{49}\phantom{00}\\
10\phantom{0}\\
\underline{7}\phantom{0}\\
30\\
\underline{28}\\
2 = \text{remainder}
\end{array}$$

## 8.4 Approximating Rationals with Decimals

Answer: $\dfrac{25{,}714}{10{,}000} = 2.5714$.

The technique used in the preceding examples can be simplified by inserting a decimal point after the last digit in the numerator and annexing zeros to the dividend in the long division process. Consider the following and Examples 2 and 5 above.

```
            .1 8 1 1                    2.5 7 1 4
    1 2 7 | 2 3.0 0 0 0          7 | 1 8.0 0 0 0
            1 2 7                        1 4
            ―――――                        ―――
            1 0 3 0                        4 0
            1 0 1 6                        3 5
            ―――――                          ―――
                1 4 0                        5 0
                1 2 7                        4 9
                ―――――                        ―――
                  1 3 0                        1 0
                  1 2 7                          7
                  ―――――                        ―――
                      3                        3 0
                                               2 8
                                               ―――
                                                 2
```

If the product of 2 and the remainder in the long division process is less than the divisor as in Examples 1, 2, and 5 above, then the *n*-place decimal obtained by the given technique is in fact the *best n*-place decimal approximation of the given rational number. If the product of 2 and the remainder is more than the divisor as in Example 3, then the last digit needs to be increased by one to get the *best n*-place decimal approximation. If the product of 2 and the remainder is equal to the divisor, then there would be two equally close decimal approximations of the given rational number. For convenience, we shall agree that the smaller will be *called* the *best n*-place decimal approximation for the number; this makes it possible to talk about *the* best *n*-place decimal approximation.* Compare the following with the preceding examples.

### Examples

1. 0.3333 is the best four-place decimal approximation of $\frac{1}{3}$.
2. 0.1811 is the best four-place decimal approximation of $\frac{23}{127}$.

---

*In physical applications, this rule is probably not the best, since it always introduces an approximation error "in the same direction." For applications, it might be much better to use the approximation (round-off) so that the last digit is even; this would tend to balance out the amounts increased and decreased in the approximation process.

3. 0.6471 is the best four-place decimal approximation of $\frac{11}{17}$.
4. 0.2462 is the best four-place decimal approximation of $\frac{197}{800}$.
5. 2.5714 is the best four-place decimal approximation of $\frac{18}{7}$.

## Exercises I

1. Which of the following are true? (a) $3.21 < 5.01$. (b) $2.023 < 2.0035$. (c) $0 < 0.005$. (d) $0.667 > \frac{2}{3}$.
2. Find the best one-place decimal approximation of each of the following rational numbers. (a) $\frac{3}{8}$. (b) $\frac{5}{13}$. (c) $5\frac{2}{7}$. (d) $6\frac{3}{5}$.
3. Find the best two-place decimal approximation of each of the following rational numbers. (a) $\frac{5}{8}$. (b) $\frac{7}{13}$. (c) $\frac{11}{15}$. (d) $\frac{3}{23}$.
4. Find the best two-place decimal approximation of each of the following rational numbers. (a) $4\frac{3}{7}$. (b) $6\frac{3}{5}$. (c) $4\frac{1}{3}$. (d) $6\frac{5}{9}$.
5. Find the best three-place decimal approximation of each of the following rational numbers. (a) $\frac{5}{8}$. (b) $\frac{3}{11}$. (c) $\frac{4}{9}$. (d) $\frac{11}{12}$.
6. Find the best three-place decimal approximation of each of the following rational numbers. (a) $4\frac{2}{5}$. (b) $5\frac{3}{11}$. (c) $6\frac{5}{6}$. (d) $2\frac{2}{9}$.
7. Find the best four-place decimal approximation of each of the following rational numbers. (a) $\frac{7}{8}$. (b) $\frac{5}{12}$. (c) $\frac{3}{127}$. (d) $\frac{23}{4,061}$.
8. Find the best four-place decimal approximation of each of the following rational numbers. (a) $5\frac{3}{23}$. (b) $6\frac{8}{17}$. (c) $2\frac{4}{19}$. (d) $3\frac{5}{13}$.
9. Find the best five-place decimal approximation of each of the following rational numbers. (a) $\frac{25}{137}$. (b) $\frac{43}{2,612}$. (c) $\frac{85}{421}$. (d) $\frac{3}{467}$.
10. Find the best five-place decimal approximation of each of the following rational numbers. (a) $4\frac{6}{11}$. (b) $1\frac{3}{5}$. (c) $4\frac{8}{9}$. (d) $3\frac{11}{413}$.

## Exercises II

1. Find the best one-place decimal approximation of each of the following rational numbers. (a) $\frac{6}{11}$. (b) $\frac{7}{12}$. (c) $3\frac{5}{8}$. (d) $6\frac{7}{15}$.

2. Find the best two-place decimal approximation of each of the following rational numbers. (a) $\frac{8}{13}$. (b) $\frac{23}{231}$. (c) $\frac{14}{71}$. (d) $\frac{6}{439}$.
3. Find the best three-place decimal approximation of each of the following rational numbers. (a) $\frac{5}{17}$. (b) $\frac{8}{25}$. (c) $\frac{11}{16}$. (d) $\frac{92}{13}$.
4. Find the best four-place decimal approximation of each of the following rational numbers. (a) $\frac{3}{23}$. (b) $\frac{4}{263}$. (c) $\frac{51}{1,206}$. (d) $\frac{29}{417}$.
5. Find the best five-place decimal approximation of each of the following rational numbers. (a) $\frac{41}{7}$. (b) $\frac{62}{321}$. (c) $\frac{3}{20}$. (d) $\frac{41}{52}$.
6. Find the best six-place decimal approximation of each of the following rational numbers. (a) $\frac{53}{128}$. (b) $\frac{62}{43}$. (c) $\frac{89}{126}$. (d) $\frac{5}{4,203}$.
7. Find the best four-place decimal approximation of each of the following rational numbers. (a) $\frac{2}{3}$. (b) $\frac{1}{2}$. (c) $\frac{823}{971}$. (d) $\frac{6,425}{8,211}$.
8. Express the sum $\frac{2}{3} + \frac{1}{2} + \frac{823}{971} + \frac{6,425}{8,211}$ as a fraction.
9. Find the best four-place decimal approximation of the sum given in Exercise 8.
10. Add the decimal approximations given as answers in Exercise 7. Compare this answer with the answer in Exercise 9.

## 8.5 Supplementary Exercises

In each of the Exercises 1 through 6, perform the indicated operations and express your answer as a proper fraction or in mixed numeral notation.
1. $(\frac{3}{8} + \frac{7}{5}) - (\frac{1}{2} - \frac{2}{5})$.
2. $(\frac{4}{9} + \frac{5}{6}) - (\frac{2}{3} + \frac{5}{18})$.
3. $6\frac{2}{3} + 8\frac{1}{2} + 5\frac{3}{8}$.
4. $7\frac{1}{2} + 2\frac{5}{6} + 9\frac{5}{12}$.
5. $3.6 + 5.2 + 9.6$.
6. $4.11 + 8.12 + 7.02$.

In Exercises 7 through 10, find the best four-place decimal approximation of the given rational numbers.
7. $\frac{17}{27}$.  8. $\frac{87}{14}$.  9. $\frac{381}{423}$.  10. $\frac{847}{23}$.

$$u \div v = \frac{u}{1} \div \frac{v}{1}$$
$$= \frac{u}{1} \times \frac{1}{v}$$
$$= \frac{u \times 1}{1 \times v}$$
$$= \frac{u}{v}$$

$$.04^{3}/_{8}$$
$$+.78^{1}/_{3}$$

$$\frac{m}{n} \times \frac{p}{q} = \frac{p}{q} \times \frac{m}{n}$$

$$84 \div 0.04\tfrac{2}{3} = 84 \div \tfrac{14/3}{100}$$
$$= 84 \div \tfrac{14}{300}$$
$$= 84 \div \tfrac{300}{14}$$
$$= 6 \times 300$$
$$=$$

# MULTIPLICATION AND DIVISION OF RATIONAL NUMBERS

## 9.1 Fractions

As we have seen, numbers are used not only for linear measure but also for measure of area. Although the methods vary for associating numbers with different types of regions such as rectangles, triangles, and circles, at this time we consider only the method used to assign a number called *area* to rectangular regions.

If one side of a rectangle has a length of 3 units and the other side has a length of 2 units, then there are $3 \times 2 = 6$ square regions one unit on a side contained in this region. We say that the area is 6 square units. (See Figure 1.) Similarly, if a rectangle has length $m$ units and width $n$ units, then the area of the rectangle is $m \times n$ square units.

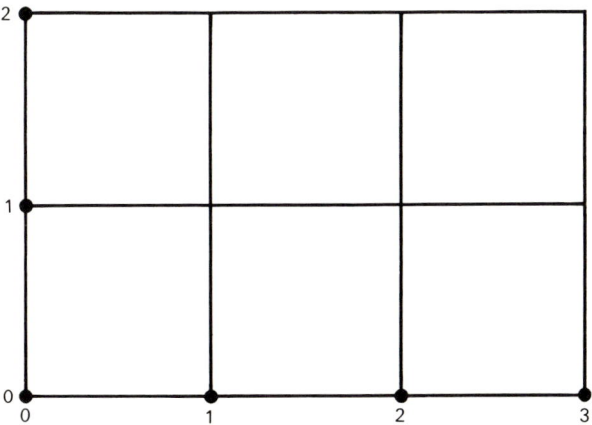

Figure 1.  $3 \times 2 = 6$ *square units*.

We use a square with each side having a length of one unit as our unit measure for area. Now, let us divide one side of this unit square into 3 equal segments and the other side into 2 equal segments and construct rectangles as in Figure 2. Since the shaded region in Figure 2 includes 2 of the 6 rectangular regions of equal size, by our interpretation of rational numbers the area of the shaded region would be $\frac{2}{6}$ of a square unit. As a natural extension of the method for assigning area given above, if the area of this rectangle with length $\frac{2}{3}$ and width $\frac{1}{2}$ is to be the product $\frac{2}{3} \times \frac{1}{2}$ then we should *define*

$$\frac{2}{3} \times \frac{1}{2} \text{ to be } \frac{2}{6}.$$

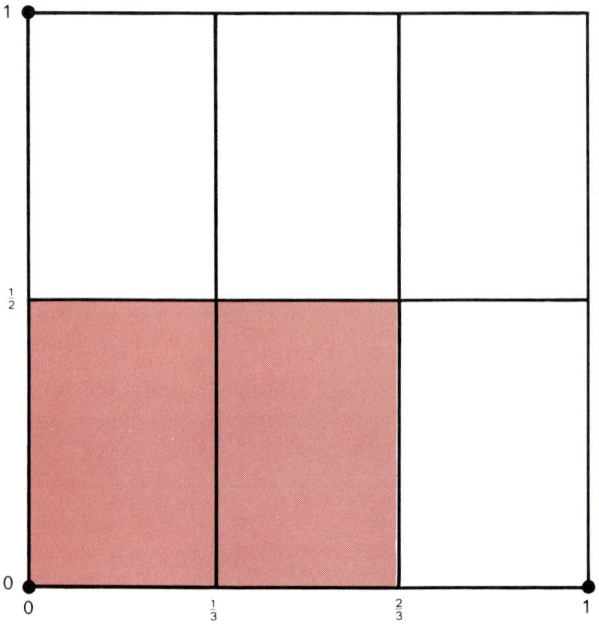

Figure 2. $\frac{2}{3} \times \frac{1}{2} = \frac{2}{6}$.

Using the above example where $\frac{2}{3} \times \frac{1}{2} = \frac{2}{6}$ and other similar examples, we are led to *define* the product of two rationals $\frac{m}{n}$ and $\frac{p}{q}$ expressed in fractional notation as follows:

$$\frac{m}{n} \times \frac{p}{q} = \frac{mp}{nq}.$$

## 9.1 Fractions

**Examples**

1. $\dfrac{3}{5} \times \dfrac{1}{2} = \dfrac{3}{10}.$  2. $\dfrac{4}{7} \times \dfrac{2}{9} = \dfrac{8}{63}.$

3. $\dfrac{3}{5} \times \dfrac{7}{8} = \dfrac{21}{40}.$  4. $\dfrac{4}{9} \times \dfrac{5}{3} = \dfrac{20}{27}.$

5. $\dfrac{5}{7} \times 0 = \dfrac{5}{7} \times \dfrac{0}{1} = \dfrac{5 \times 0}{7 \times 1} = \dfrac{0}{7} = 0.$

It is an immediate consequence of the associative and commutative properties of multiplication for whole numbers that the *set of rational numbers have the associative and commutative properties of multiplication*. That is,

$$\frac{m}{n} \times \frac{p}{q} = \frac{p}{q} \times \frac{m}{n},$$

and

$$\frac{m}{n} \times \left(\frac{p}{q} \times \frac{r}{s}\right) = \left(\frac{m}{n} \times \frac{p}{q}\right) \times \frac{r}{s}.$$

Notice that $\frac{x}{y} \times \frac{y}{x} = \frac{xy}{yx} = 1$. We call $\frac{y}{x}$ the **multiplicative inverse** of $\frac{x}{y}$ since $\frac{y}{x}$ is the rational number whose product with the rational $\frac{x}{y}$ is the multiplicative identity 1. (Of course, we assume neither rational is zero.) The *fraction* "$\frac{y}{x}$" is called the **reciprocal** of "$\frac{x}{y}$". Observe that the multiplicative inverse of $\frac{x}{y}$ is a rational number; the term "reciprocal" refers to a notation (fraction) for the multiplicative inverse.

We can find the product $\frac{20}{21} \times \frac{27}{8}$ and express it as a fraction in lowest terms as follows.

$$\frac{20}{21} \times \frac{27}{8} = \frac{540}{168}$$

$$= \frac{12 \times 45}{12 \times 14}$$

$$= \frac{45}{14}.$$

Another procedure to find the product $\frac{20}{21} \times \frac{27}{8}$ can be obtained by using the definition of multiplication, the fact that $\frac{x}{y} \times \frac{y}{x} = 1$, and the associative and commutative properties of multiplication. It is indicated as follows:

$$\frac{20}{21} \times \frac{27}{8} = \frac{5 \times 4}{7 \times 3} \times \frac{3 \times 9}{4 \times 2}$$

$$= \frac{5}{7} \times \left(\frac{4}{3} \times \frac{3}{4}\right) \times \frac{9}{2}$$

$$= \frac{5}{7} \times \frac{9}{2}$$

$$= \frac{45}{14}.$$

This last technique can be simplified by "cancelling" equal *factors* in the numerators and denominators of the fractions.

$$\frac{20}{21} \times \frac{27}{8} = \frac{5 \times \cancel{4}}{7 \times \cancel{3}} \times \frac{\cancel{3} \times 9}{\cancel{4} \times 2}$$

$$= \frac{45}{14}.$$

Of course, other variations of this technique can be used.

From the definition of division, $\frac{m}{n} \div \frac{p}{q}$ is the rational number $\frac{x}{y}$ such that

$$\frac{p}{q} \times \frac{x}{y} = \frac{m}{n},$$

or

$$\frac{px}{qy} = \frac{m}{n}.$$

Considering the cancelling technique described in the preceding paragraph, we see that if $x = mq$ and $y = np$, then

$$\frac{px}{qy} = \frac{p(mq)}{q(np)} = \frac{m}{n}.$$

Therefore,

$$\frac{m}{n} \div \frac{p}{q} = \frac{mq}{np}.$$

Since $\frac{mq}{np} = \frac{m}{n} \times \frac{q}{p}$, the quotient $\frac{m}{n} \div \frac{p}{q}$ is the product of $\frac{m}{n}$ and the multiplicative inverse of $\frac{p}{q}$. Sometimes, we say that *to divide with fractions we invert the divisor and multiply*. (We also say that you multiply by the reciprocal of the divisor, but here again we are taking

## 9.1 Fractions

the usual liberties of not distinguishing between the number and its name.)

**Examples**

1. $\dfrac{3}{8} \div \dfrac{4}{7} = \dfrac{3}{8} \times \dfrac{7}{4} = \dfrac{21}{32}.$

2. $0 \div \dfrac{2}{3} = 0 \times \dfrac{3}{2} = 0.$

3. $\dfrac{37}{12} \div \dfrac{13}{20} = \dfrac{37}{12} \times \dfrac{20}{13} = \dfrac{37}{3} \times \dfrac{5}{13} = \dfrac{185}{39}.$

4. $\dfrac{2}{9} \div 6 = \dfrac{2}{9} \times \dfrac{1}{6} = \dfrac{1}{9} \times \dfrac{1}{3} = \dfrac{1}{27}.$

As with whole numbers, we cannot divide by zero. Thus, $\frac{3}{7} \div 0$ and $\frac{7}{9} \div \frac{0}{5}$, for example, are undefined. However, it is important to observe that *the set consisting of all rational numbers different from zero is closed with respect to division.* Therefore, we finally have a set of numbers which is closed with respect to addition, multiplication, and division. In the next chapter, we extend the number system once more and obtain closure with respect to all four rational operations, except division by zero.

One can derive techniques to find such products as $46\frac{1}{3} \times 42\frac{1}{2}$, or quotients such as $36\frac{1}{5} \div 27\frac{2}{3}$, using the mixed numeral notation directly. However, it is usually easier to express each rational number in fractional notation and then proceed as discussed earlier. However, for a product such as $86 \times 42\frac{1}{2}$, the following technique is quite useful.

$$\begin{array}{r} 86 \\ 42\frac{1}{2} \\ \hline 43 \\ 172 \\ 344 \\ \hline 3655 \end{array} \begin{array}{l} \\ \leftrightarrow \frac{1}{2} \times 86 \\ \Big\} \leftrightarrow \text{Sum} = 3{,}612 = 86 \times 42 \\ \\ \end{array}$$

This technique can be justified as follows.

$$\begin{aligned} 86 \times 42\tfrac{1}{2} &= 86 \times (42 + \tfrac{1}{2}) \\ &= (86 \times 42) + (86 \times \tfrac{1}{2}) \\ &= 3{,}612 + 43 \\ &= 3{,}655. \end{aligned}$$

## Exercises I

In Exercises 1 through 5, find the indicated product and express the answer as a proper fraction or in the mixed numeral notation.

1. (a) $\frac{2}{3} \times \frac{5}{7}$.  (b) $\frac{3}{11} \times \frac{5}{8}$.
   (c) $\frac{4}{7} \times \frac{3}{2}$.  (d) $\frac{9}{10} \times \frac{4}{9}$.
2. (a) $\frac{5}{12} \times \frac{4}{3}$.  (b) $\frac{7}{5} \times \frac{11}{14}$.
   (c) $\frac{6}{11} \times \frac{33}{2}$.  (d) $\frac{14}{15} \times \frac{3}{2}$.
3. (a) $\frac{12}{7} \times \frac{21}{32}$.  (b) $\frac{9}{10} \times \frac{4}{3}$.
   (c) $\frac{13}{17} \times \frac{21}{26}$.  (d) $\frac{14}{11} \times \frac{18}{14}$.
4. (a) $\frac{6}{5} \times \frac{7}{5}$.  (b) $\frac{2}{7} \times \frac{63}{14}$.
   (c) $\frac{4}{11} \times \frac{20}{17}$.  (d) $\frac{16}{11} \times \frac{44}{53}$.
5. (a) $\frac{2}{11} \times \frac{5}{11}$.  (b) $\frac{16}{3} \times \frac{20}{61}$.
   (c) $\frac{4}{5} \times \frac{3}{8}$.  (d) $\frac{15}{29} \times \frac{58}{5}$.

In Exercises 6 through 10, find the indicated quotient and express the answer as a proper fraction or in the mixed numeral notation.

6. (a) $\frac{2}{3} \div \frac{5}{7}$.  (b) $\frac{3}{11} \div \frac{5}{8}$.
   (c) $\frac{4}{7} \div \frac{3}{2}$.  (d) $\frac{9}{10} \div \frac{4}{9}$.
7. (a) $\frac{5}{12} \div \frac{4}{3}$.  (b) $\frac{7}{5} \div \frac{11}{14}$.
   (c) $\frac{6}{11} \div \frac{33}{2}$.  (d) $\frac{14}{15} \div \frac{3}{2}$.
8. (a) $\frac{12}{7} \div \frac{21}{32}$.  (b) $\frac{9}{10} \div \frac{4}{3}$.
   (c) $\frac{13}{17} \div \frac{21}{26}$.  (d) $\frac{14}{11} \div \frac{18}{14}$.
9. (a) $\frac{6}{5} \div \frac{7}{5}$.  (b) $\frac{2}{7} \div \frac{63}{14}$.
   (c) $\frac{4}{11} \div \frac{20}{17}$.  (d) $\frac{16}{11} \div \frac{44}{53}$.
10. (a) $\frac{2}{11} \div \frac{5}{11}$.  (b) $\frac{16}{3} \div \frac{20}{61}$.
    (c) $\frac{4}{5} \div \frac{3}{8}$.  (d) $\frac{15}{29} \div \frac{58}{5}$.

## Exercises II

1. Give reasons for the steps in the proof that $u \div v = \frac{u}{v}$, provided $v \neq 0$.

   Step 1. $\qquad u \div v = \frac{u}{1} \div \frac{v}{1}$.

   Step 2. $\qquad\qquad\quad = \frac{u}{1} \times \frac{1}{v}$.

   Step 3. $\qquad\qquad\quad = \frac{u \times 1}{1 \times v}$.

   Step 4. $\qquad\qquad\quad = \frac{u}{v}$.

## 9.2 Decimals

In Exercises 2 through 9, perform the indicated operations and express your answer as a proper fraction or in the mixed numeral notation.

2. $(\frac{2}{3} \times \frac{4}{7}) + \frac{5}{8}$.
3. $\frac{2}{3}(\frac{4}{9} + \frac{5}{6}) + \frac{5}{2}$.
4. $\frac{5}{7} \times (\frac{3}{8} \div \frac{4}{9})$.
5. $(\frac{5}{7} \times \frac{3}{8}) \div \frac{4}{9}$.
6. $(\frac{6}{11} - \frac{1}{3}) \times (\frac{4}{7} + \frac{2}{3})$.
7. $\frac{5}{9}(\frac{4}{5} + \frac{2}{3}) - (\frac{2}{3} \div \frac{5}{2})$.
8. $(\frac{5}{9} \times \frac{4}{15}) \div (\frac{3}{8} \times \frac{5}{2})$.
9. $(\frac{5}{6} + \frac{7}{12}) + (\frac{11}{10} \div \frac{121}{40})$.
10. In Section 6.1, page 87, we let $\phi(n)$ represent the number of natural numbers less than $n$ which are relatively prime to $n$. A formula known as *Euler's phi-function* can be used to determine $\phi(n)$. If $p_1, p_2, p_3, \ldots, p_k$ are the prime factors of $n$, then

$$\phi(n) = n\left(1 - \frac{1}{p_1}\right)\left(1 - \frac{1}{p_2}\right)\left(1 - \frac{1}{p_3}\right) \times \cdots \times \left(1 - \frac{1}{p_k}\right).$$

Use this formula to find each of the following.
 (a) $\phi(3)$.    (b) $\phi(5)$.    (c) $\phi(7)$.
 (d) $\phi(12)$.   (e) $\phi(15)$.   (f) $\phi(28)$.
 (g) $\phi(35)$.   (h) $\phi(40)$.   (i) $\phi(60)$.


## 9.2 Decimals

Let us determine techniques to find the product of two rationals such as 4.81 and 5.213 using the decimal notation. Since

$$4.81 = \frac{481}{100} \quad \text{and} \quad 5.213 = \frac{5{,}213}{1{,}000},$$

$$4.81 \times 5.213 = \frac{481}{100} \times \frac{5{,}213}{1{,}000}$$

$$= \frac{481 \times 5{,}213}{100 \times 1{,}000}$$

$$= \frac{2{,}507{,}453}{100{,}000}$$

$$= 25.07453.$$

Essentially, what is involved in this procedure is to determine the product 481 × 5,213 by the multiplication algorithm and to place the decimal point. Since the two-place decimal 4.81 can be written as a decimal fraction with $10^2$ as denominator and since 5.213 can be written as a decimal fraction with $10^3$ as denominator, the product can be written as a decimal fraction with $10^2 \times 10^3 = 10^5$ as denominator. Hence, the decimal notation for the product will be a five-place decimal. In general, the product of an $m$-place decimal and an $n$-place decimal will be an $(m + n)$-place decimal.

## Examples

1.  
```
      3.2 1 7   (3-place)
        4 3.2   (1-place)
      6 4 3 4
    9 6 5 1
  1 2 8 6 8
  1 3 8.9 7 4 4   (4-place)
```

2.  
```
    4.5 1 2 3   (4-place)
    0.2 1 5     (3-place)
    2 2 5 6 1 5
    4 5 1 2 3
    9 0 2 4 6
    1.0 7 0 1 4 4 5   (7-place)
```

To find the quotient $130.572 \div 0.17$, we need to find the number $x$ such that $0.17x = 130.572$. Since

$$130.572 \div 0.17 = \frac{130{,}572}{1{,}000} \div \frac{17}{100}$$

$$= \frac{130{,}572}{1{,}000} \times \frac{100}{17}$$

$$= \frac{130{,}572}{170},$$

the quotient is the rational number $\frac{130{,}572}{170}$. It should be obvious that the quotient of any two numbers written in decimal notation is a rational number; however, the quotient may not have a simple terminating decimal representation. For example, $\frac{1}{6} \div \frac{1}{2} = \frac{1}{6} \times \frac{2}{1} = \frac{1}{3}$, and $\frac{1}{3}$ cannot be expressed as a finite decimal.

In Section 8.4, we found that any rational number can be approximated by a rational number written in decimal notation and that the approximation can be obtained by using the long division algorithm. For example, to find a one-place decimal approximation of the quotient $\frac{130{,}572}{170}$ which is less than or equal to $\frac{130{,}572}{170}$, we seek the greatest integer $x$ such that $\frac{x}{10} \leq \frac{130{,}572}{170}$, or $17x \leq 130{,}572$.

## 9.2 Decimals

```
         7,6 8 0
    17 ) 1 3 0,5 7 2
         1 1 9
         —————
           1 1 5
           1 0 2
           —————
             1 3 7
             1 3 6
             —————
               1 2  = remainder
```

Thus, 7,680 is the greatest integer $x$ such that $\frac{x}{10} \leq \frac{130,572}{170}$; therefore, $\frac{7,680}{10} = 768.0$ is the best one-place decimal approximation which is less than or equal to the quotient $130.572 \div 0.17$. Since 2 times the remainder is greater than the divisor, we conclude that 768.1 is the best one-place decimal approximation of the quotient.

Since $130.572 \div 0.17 = (130.572)(100) \div (0.17)(100)$, we can say that $130.572 \div 0.17 = 13,057.2 \div 17$. Thus, the quotient of the two numbers is equal to the quotient obtained by dividing 13,057.2 by 17. The usual method of finding the one-place decimal approximation of the quotient is the following.

```
          7 6 8.0
    .1 7 ) 1 3 0.5 7 2
           1 1 9
           —————
             1 1 5
             1 0 2
             —————
               1 3 7
               1 3 6
               —————
                 1 2
```

*Explanation of the Procedure.* To find the decimal approximation of the quotient of two numbers where the divisor is written as an $n$-place decimal, we multiply the divisor and dividend by $10^n$ (that is, "move" the decimals $n$-places to the right) so that the divisor will be an integer.

### Examples

Find the best three-place decimal approximation of each of the following.

(a) 68.21743 ÷ 4.23

```
            1 6.1 2 7
4.2 3 ) 6 8.2 1 7 4 3
        4 2 3
        ─────
        2 5 9 1
        2 5 3 8
        ───────
          5 3 7
          4 2 3
          ─────
          1 1 4 4
            8 4 6
          ───────
            2 9 8 3
            2 9 6 1
            ───────
                2 2
```

Answer: 16.127

(b) 123.15 ÷ 18.3

```
              6.7 8 4
18.3 ) 1 2 3.1 5 0 0
       1 0 8 8
       ───────
         1 4 3 5
         1 2 8 1
         ───────
           1 5 4 0
           1 4 6 4
           ───────
               7 6 0
               7 3 2
               ─────
                 2 8
```

Answer: 6.784

### Exercises I

Find the indicated products and express your answer in decimal notation.

1. 28.3 × 17.41.
2. 3.81 × 0.012.
3. 1.413 × 3.107.
4. 4.2865 × 0.0031.
5. 4.81 × 0.0085.
6. 0.0612 × 0.0028.

Find the best three-place decimal approximation of each of the following quotients.

7. 7.2346 ÷ 3.8.
8. 4.371 ÷ 2.51.
9. 8.13 ÷ 2.614.
10. 0.02715 ÷ 0.0059.

### Exercises II

Find the indicated products and express your answer in decimal notation.

1. 37.3 × 45.003.
2. 8.31 × 0.0023.
3. 5.011 × 0.0301.
4. 8.964 × 24.107.
5. 52.3 × 0.00052.
6. 0.0023 × 0.000671.

Find the best five-place decimal approximation of each of the following quotients.

## 9.3 Complex Fractions and Complex Decimals

7. $3.954 \div 0.23$.
8. $3.01125 \div 4.37$.
9. $0.0003712 \div 0.239$.
10. $47.053 \div 9.15$.

## 9.3 Complex Fractions and Complex Decimals

In the fractional notation, we have insisted that the numerator and denominator be whole numbers. It is occasionally convenient to use the notation $\frac{m/n}{p/q}$ to represent the quotient $\frac{m}{n} \div \frac{p}{q}$. A fraction where the numerator or denominator is a fraction is called a complex fraction.

To express the rational $\frac{2/3 + 5/7}{5/6 + 3/2}$ as a fraction in lowest terms, we can proceed as follows.

$$\frac{\frac{2}{3} + \frac{5}{7}}{\frac{5}{6} + \frac{3}{2}} = \frac{\frac{14}{21} + \frac{15}{21}}{\frac{5}{6} + \frac{9}{6}}$$

$$= \frac{\frac{29}{21}}{\frac{14}{6}}$$

$$= \frac{29}{21} \div \frac{14}{6}$$

$$= \frac{29}{21} \times \frac{6}{14}$$

$$= \frac{29}{\cancel{3} \times 7} \times \frac{\cancel{3} \times \cancel{2}}{7 \times \cancel{2}}$$

$$= \frac{29}{49}$$

Another and usually more efficient approach is to multiply the numerator and denominator of the complex fraction by the L.C.D. of all the fractions; in this example, the L.C.D. is 42. This technique is demonstrated as follows.

$$\frac{\frac{2}{3} + \frac{5}{7}}{\frac{5}{6} + \frac{3}{2}} = \frac{42(\frac{2}{3} + \frac{5}{7})}{42(\frac{5}{6} + \frac{3}{2})}$$

$$= \frac{28 + 30}{35 + 63}$$

$$= \frac{58}{98}$$

$$= \frac{2 \times 29}{2 \times 49}$$

$$= \frac{29}{49}.$$

We found in Section 8.4 that $\frac{1}{3}$ cannot be expressed as an *n*-place decimal. However, $\frac{1}{3}$ can be denoted, for example, by "$0.33\frac{1}{3}$"; this is called a **complex decimal**. The complex decimal $0.33\frac{1}{3}$ represents the sum

$$\frac{3}{10} + \frac{3\frac{1}{3}}{100}.$$

Notice that

$$\frac{3}{10} + \frac{3\frac{1}{3}}{100} = \frac{3}{10} + \frac{\frac{10}{3}}{100}$$

$$= \frac{3}{10} + \left(\frac{10}{3} \div 100\right)$$

$$= \frac{3}{10} + \left(\frac{10}{3} \times \frac{1}{100}\right)$$

$$= \frac{3}{10} + \frac{1}{30}$$

$$= \frac{9}{30} + \frac{1}{30}$$

$$= \frac{10}{30}$$

$$= \frac{1}{3}.$$

Complex decimals are encountered quite often; it is not unusual to find an item priced at $0.87\frac{1}{2}$ or to find interest rates quoted as $0.05\frac{1}{4}$.

If we clearly understand the complex decimal and mixed numeral notations, the techniques required for performing the arithmetic operations should be rather obvious. Consider the following examples.

## 9.3 Complex Fractions and Complex Decimals

### Examples

1. Add: $\quad 0.05\frac{1}{4} = 0.05\frac{2}{8}$
   $\qquad\qquad 0.06\frac{3}{8} = 0.06\frac{3}{8}$
   $\qquad\qquad \overline{\qquad\quad\ \ 0.11\frac{5}{8}}$

2. Add: $\quad 0.03\frac{1}{3} = 0.03\frac{1}{3}$
   $\qquad\qquad 0.4\frac{1}{2} = 0.45$
   $\qquad\qquad \overline{\qquad\quad\ \ 0.48\frac{1}{3}}$

   Note: $4\frac{1}{2} \div 10 = 45 \div 100$

3. Multiply:

   $\quad\ \ 4\ 6\ 2$
   $\ 0.0\ 3\frac{1}{3}$
   $\overline{\quad 1\ 5\ 4} = \frac{1}{3} \times 462$
   $\ 1\ 3\ 8\ 6\ \ = 3 \times 462$
   $\overline{1\ 5.4\ 0}$

4. Multiply:

   $\quad\ \ 5\ 2\ 1$
   $\ .0\ 0\ 4\frac{1}{3}$
   $\overline{\quad 1\ 7\ 3\frac{2}{3}} = \frac{1}{3} \times 521$
   $\ 2\ 0\ 8\ 4\ \ = 4 \times 521$
   $\overline{2.2\ 5\ 7\frac{2}{3}}$

5. Divide: $\quad 84 \div 0.04\frac{2}{3} = 84 \div \dfrac{14\frac{2}{3}}{100}$

   $\qquad\qquad\qquad\qquad\ = 84 \div \dfrac{14}{300}$

   $\qquad\qquad\qquad\qquad\ = 84 \times \dfrac{300}{14}$

   $\qquad\qquad\qquad\qquad\ = 6 \times 300$

   $\qquad\qquad\qquad\qquad\ = 1{,}800.$

### Exercises I

1. Find the sum and express your answer in complex decimal notation.

   (a) $0.12\frac{1}{2}$
   $\quad\ \ 0.23\frac{2}{3}$

   (b) $0.04\frac{1}{5}$
   $\quad\ \ 0.11\frac{2}{9}$

   (c) $0.08\frac{2}{5}$
   $\quad\ \ 0.007\frac{5}{7}$

2. Find the sum and express your answer in complex decimal notation.

   (a) $0.21\frac{1}{3}$
   $\quad\ \ 0.46\frac{2}{5}$

   (b) $1.28\frac{1}{2}$
   $\quad\ \ 0.21\frac{1}{7}$

   (c) $0.38\frac{2}{3}$
   $\quad\ \ 4.12\frac{1}{6}$

3. Find the difference and express your answer in complex decimal notation.

(a) $0.23\frac{2}{3}$ \qquad (b) $0.14\frac{2}{3}$ \qquad (c) $0.08\frac{2}{5}$
$\underline{0.18\frac{1}{2}}$ \qquad\quad $\underline{0.04\frac{4}{5}}$ \qquad\quad $\underline{0.007\frac{5}{7}}$

4. Find the difference and express your answer in complex decimal notation.

    (a) $0.32\frac{1}{5}$ \qquad (b) $3.16\frac{1}{2}$ \qquad (c) $4.12\frac{1}{6}$
    $\underline{0.21\frac{2}{3}}$ \qquad\quad $\underline{0.32\frac{1}{6}}$ \qquad\quad $\underline{0.52\frac{2}{3}}$

5. Find the product and express your answer in complex decimal notation.

    (a) $864$ \qquad (b) $425$ \qquad (c) $356$
    $\underline{.04\frac{1}{3}}$ \qquad\quad $\underline{.21\frac{1}{5}}$ \qquad\quad $\underline{.02\frac{1}{2}}$

6. Find the product and express your answer in complex decimal notation.

    (a) $428$ \qquad (b) $803$ \qquad (c) $268$
    $\underline{.05\frac{1}{4}}$ \qquad\quad $\underline{.86\frac{1}{2}}$ \qquad\quad $\underline{2.7\frac{3}{8}}$

7. Find the quotients and express your answer in complex decimal notation.

    (a) $275 \div 0.02\frac{1}{2}$. \quad (b) $6.4 \div 0.04\frac{4}{7}$. \quad (c) $84 \div 0.002\frac{1}{3}$.

In Exercises 8 through 10, perform the indicated operations and express your answer as a proper fraction or in mixed numeral notation.

8. (a) $\dfrac{\frac{2}{5}+\frac{4}{3}}{\frac{5}{6}+\frac{1}{2}}$. \qquad (b) $\dfrac{\frac{7}{8}+\frac{5}{6}}{\frac{3}{2}+\frac{5}{12}}$.

9. (a) $\dfrac{\frac{4}{7}+\frac{1}{6}}{\frac{5}{21}+\frac{7}{2}}$. \qquad (b) $\dfrac{\frac{7}{20}+\frac{3}{10}}{\frac{5}{2}+\frac{8}{5}}$.

10. (a) $\dfrac{\frac{5}{9}+\frac{5}{4}}{\frac{7}{18}+\frac{11}{3}}$. \qquad (b) $\dfrac{\frac{7}{30}+\frac{8}{15}}{\frac{9}{4}+\frac{11}{6}}$.

## Exercises II

In Exercises 1 through 10, perform the indicated operations.

1. $3.14\frac{2}{3} + 0.86\frac{1}{2}$. \qquad 2. $2.02\frac{1}{6} + 0.04\frac{2}{5}$.
3. $5.12\frac{3}{8} - 2.24\frac{1}{5}$. \qquad 4. $0.08\frac{1}{3} - 0.05\frac{1}{4}$.
5. $0.05\frac{1}{4} \times 356$. \qquad\quad 6. $0.082\frac{1}{3} \times 300$.
7. $\dfrac{\frac{2}{3}+\frac{7}{5}}{\frac{3}{8}+\frac{9}{2}}$. \qquad 8. $\dfrac{\frac{5}{7}+\frac{3}{10}}{\frac{8}{5}+\frac{17}{35}}$.

9. $\dfrac{\frac{4}{9}+\frac{5}{12}}{\frac{1}{2}+\frac{5}{18}}$.

10. $\dfrac{\frac{3}{11}+\frac{7}{4}}{\frac{5}{22}+\frac{5}{44}}$.

## 9.4 Supplementary Exercises

In Exercises 1 through 6, perform the indicated operations.
1. $2.314 + 0.012 + 4.23 + 5.7611 + 5.01$.
2. $0.0231 \times 684.3$.
3. $2.1395 \times 23.56$.
4. $(5.67 + 3.1) \times 0.2041$.
5. $(51.32 + 3.51) \times (2.7 + 3.24)$.
6. $(53.41 - 32.785) \times 30.01$.

In Exercises 7 through 10, find the best three-place decimal approximation of the answer.

7. $41.3 \div 3.071$.

8. $0.318 \div 5.3$.

9. $\dfrac{\frac{5}{3}+\frac{3}{7}}{\frac{2}{7}+\frac{8}{21}}$.

10. $\dfrac{\frac{5}{2}+\frac{11}{17}}{\frac{3}{2}+\frac{15}{34}}$.

# 10

$= -(b - a)$

$(a)(-b) = (-b)(a) = -(ab)$

```
         0  1  2  3  4  5  6  7  8
|--|--|--|--|--|--|--|--|--|--|--|--|--|--|--|--|
-8 -7 -6 -5 -4 -3 -2 -1  0
```

If $b > a$, then $a + (-b)$

$2 + (-2) = 0$

$[(-\tfrac{1}{2})(\tfrac{4}{5})] \div [(\tfrac{-1}{3}) + (\tfrac{2}{-5})]$

# THE NEGATIVE INTEGERS AND RATIONALS

## 10.1 Extending the Number Line

The next extension we make of the number system is the development of the integers. Technically, the set of positive integers, which we occasionally denote by +1, +2, +3, +4, . . . (or $^+1$, $^+2$, $^+3$, $^+4$, . . .), and the set of natural numbers 1, 2, 3, 4, . . . are different sets. However, the two sets of numbers with the operations of addition and multiplication are systems which mathematicians call *isomorphic* (same form). Roughly speaking, these sets are indistinguishable; it is like having two different names for the elements in a given set. Although we did not emphasize this earlier, the same situation exists with the set of natural numbers 1, 2, 3, 4, . . . and the set of rationals $\frac{1}{1}, \frac{2}{1}, \frac{3}{1}, \frac{4}{1}, \ldots$ . In doing arithmetic calculations, either notation could be used with the same results.

Henceforth, we shall use "natural numbers" and "positive integers" interchangeably and consider both terms as naming the set whose elements are 1, 2, 3, 4, . . . . the new numbers which we create are the negative integers; these are denoted by −1, −2, −3, −4, . . . (or $^-1$, $^-2$, $^-3$, $^-4$, . . .). The negative integer "−2" (read "negative two"), for example, is called the opposite, or additive inverse, of 2. We shall define addition in such a way that

$$2 + (-2) = 0;$$

that is, the sum of a positive integer and its additive inverse is the additive identity zero. * For this reason, 2 is also called the additive inverse of −2.

---

*We should point out that one advantage of the second notation given for the negative integers is that we may write 2 + $^-$2 = 0 and avoid parentheses without unnecessary confusion. However, we use the more traditional notation which is also standard in advanced mathematics.

In terms of the number line, we have extended the number line to the left of the origin (zero). As seen in Figure 1, the negative integer −1 is paired with the end point of the first segment of unit length to the left of zero; the negative integer −2 is paired with the end point of the second such segment to the left of zero; the negative integer −3 is paired with the end point of the third segment of unit length to the left of zero; etc. The set consisting of the positive integers, the negative integer and zero is called the set of *integers*.

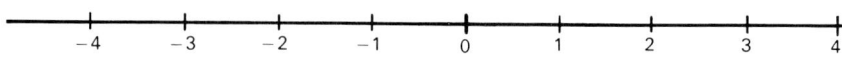

Figure 1. *The integers.*

After one constructs (invents) a number system, there are generally two reasons for considering an extension. One is the use or applications which the new numbers have. The other reason is more sophisticated; essentially, it is to have a more complete mathematical structure. For example, with the positive integers we have closure with respect to addition and multiplication but not with respect to subtraction or division. Enlarging the number system to include the positive rationals, we obtain closure with respect to division, except division by zero. Now, after we extend our number system to include the negative numbers we shall have closure with respect to all the rational operations, except division by zero. In Chapter 12, when we introduce the algebraic operation of taking roots, we shall again be motivated to extend the number system to obtain closure.

When deciding how to *define* addition and multiplication for the set of integers, we should not lose sight of the fact that the set of positive integers is a subset of the set of integers. Therefore, we want our definitions of addition and multiplication for the set of integers to be consistent with the definitions for the set of positive integers. In addition, we want our definitions to preserve, if possible, all the additive and multiplicative properties for the set of positive integers. Let us see how this is accomplished.

We have significant interpretations of addition and subtraction in terms of the number line. The sum 8 + 3 is the number associated with the point 3 units to the *right* of 8. The sum 3 + 8 is the number associated with the point 8 units to the right of 3; in both cases, the number is 11. The difference 8 − 3 is the number

## 10.1 Extending the Number Line

associated with the point 3 units to the *left* of 8; it is 5. Thus, to be consistent with this interpretation, 3 − 8 should be the number associated with the point 8 units to the left of 3; the number is −5. (See Figure 2.)

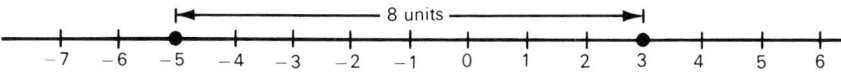

Figure 2. 3 − 8 = −5.

If 3 − 8 = −5, then by the definition of subtraction 8 + (−5) = 3. Thus, to *add* −5 to 8 we move 5 units to the *left* of 8. We also see that if the sum (−5) + 8 is the number associated with the point 8 units to the right of −5, then (−5) + 8 = 3, and the commutative property of addition is preserved since 8 + (−5) = (−5) + 8.

By our previous interpretations, we observe that 7 − 2 and 7 + (−2) is the same number. Similarly, 3 − 10 and 3 + (−10) is the same number. Also, we conclude that (−5) − 6 and (−5) + (−6) is −11.

Finally, consider the sum (−7) − (−8). By the definition of subtraction, the difference (−7) − (−8) should be a number $x$ such that

$$(-8) + x = -7.$$

If we start at the point paired with −8, then to get to −7 we need to move 1 unit to the *right*; thus, $x$ should be 1. Consequently,

$$(-7) - (-8) = 1,$$

which, of course, is the sum (−7) + 8.

The following examples should make the rules (definitions) for addition and subtraction quite obvious.

### Examples

1. Add:

| −3 | −8  | −3 | −5  | −7  |
|----|-----|----|-----|-----|
| −8 | −10 | −6 | −11 | −7  |
| −11| −18 | −9 | −16 | −14 |

2. Add:

| −3 | −7 | 3  | 10 | 6  |
|----|----|----|----|----|
| 8  | 10 | −8 | −7 | −6 |
| 5  | 3  | −5 | 3  | 0  |

3. Subtract:

$$\begin{array}{r}-3\\-8\\\hline 5\end{array}\qquad\begin{array}{r}-8\\-10\\\hline 2\end{array}\qquad\begin{array}{r}-6\\-3\\\hline -3\end{array}\qquad\begin{array}{r}-10\\-3\\\hline -7\end{array}\qquad\begin{array}{r}-8\\-8\\\hline 0\end{array}$$

4. Subtract:

$$\begin{array}{r}-3\\8\\\hline -11\end{array}\qquad\begin{array}{r}-8\\10\\\hline -18\end{array}\qquad\begin{array}{r}-6\\3\\\hline -9\end{array}\qquad\begin{array}{r}-10\\3\\\hline -13\end{array}\qquad\begin{array}{r}-8\\8\\\hline -16\end{array}$$

5. Subtract:

$$\begin{array}{r}3\\-8\\\hline 11\end{array}\qquad\begin{array}{r}8\\-10\\\hline 18\end{array}\qquad\begin{array}{r}6\\-3\\\hline 9\end{array}\qquad\begin{array}{r}10\\-3\\\hline 13\end{array}\qquad\begin{array}{r}8\\-8\\\hline 16\end{array}$$

We extend the number system in what should now be an obvious manner. For each positive rational number $x$ we postulate the existence of a rational called the additive inverse of $x$ which is denoted by $-x$. For example, $-\frac{3}{4}$ (read "negative three-fourths") is the additive inverse of $\frac{3}{4}$. The same rules for addition and subtraction apply to the set of rationals as for the set of integers. Let us state them explicitly.

*Let a and b be positive rational numbers.*

**Rule 1.** $(-a) + 0 = 0 + (-a) = -a.$

**Rule 2.** $a + (-a) = (-a) + a = 0.$

**Rule 3.** $(-a) - 0 = -a.$

**Rule 4.** $(-a) - (-a) = 0.$

**Rule 5.** $(-a) + (-b) = -(a + b).$

**Rule 6.** If $a > b$, then $a + (-b) = a - b.$

**Rule 7.** If $b > a$, then $a + (-b) = -(b - a).$

**Rule 8.** $a - (-b) = a + b.$

**Rule 9.** $(-a) - (-b) = (-a) + b.$

**Rule 10.** $(-a) - b = -(a + b).$

## 10.1 Extending the Number Line

### Exercises I

1. Add:  (a) $-17$ $\phantom{+}12$  (b) $-18$ $-35$  (c) $-23$ $\phantom{+}37$

2. Add:  (a) $-11$ $\phantom{+}15$  (b) $-37$ $-17$  (c) $-322$ $\phantom{+}187$

3. Add:  (a) $-16$ $\phantom{+}14$  (b) $-23$ $\phantom{+}39$  (c) $-38$ $\phantom{+}38$

4. (a) $\frac{3}{4} + (-\frac{7}{8})$. (b) $(-\frac{11}{6}) + (-\frac{2}{3})$. (c) $\frac{1}{2} + (-\frac{4}{7})$.
5. (a) $(-\frac{11}{16}) + \frac{1}{4}$. (b) $(-\frac{1}{5}) + \frac{3}{8}$. (c) $(-\frac{3}{8}) + (-\frac{2}{9})$.
6. Subtract:
   (a) $-18$ $\phantom{+}12$  (b) $-42$ $-17$  (c) $-38$ $-92$

7. Subtract:
   (a) $-16$ $\phantom{+}25$  (b) $-23$ $-58$  (c) $47$ $93$

8. Subtract:
   (a) $-17$ $-17$  (b) $-145$ $-378$  (c) $-576$ $-398$

9. (a) $(-\frac{5}{6}) - \frac{4}{7}$. (b) $(-\frac{3}{8}) - (-\frac{5}{9})$. (c) $\frac{5}{12} - (-\frac{11}{16})$.
10. (a) $\frac{4}{5} - \frac{9}{14}$. (b) $(-\frac{9}{11}) - (-\frac{3}{4})$. (c) $(-\frac{7}{8}) - \frac{2}{3}$.
11. (a) $3.572 - 5.376$. (b) $(-5.21) + (3.78)$. (c) $(-8.3) - (-9.1)$.
12. (a) $(4.23) - (-4.81)$. (b) $(-11.3) - (-7.4)$. (c) $(-2.05) - (-3.27)$.

### Exercises II

1. If $a = -3$, $b = -7$, and $c = 8$, show that $a + (b + c) = (a + b) + c$.
2. If $a = -9$, $b = -6$, and $c = -3$, show that $a + (b + c) = (a + b) + c$.
3. If $a = -8$, $b = 4$, and $c = -11$, show that $a + (b + c) = (a + b) + c$.

In Exercises 4 through 10, perform the indicated operations.
4. $[(-11) - (-23)] - [(-34) + (-15)]$.
5. $[(-47) - (-23)] + [(-15) - 72]$.
6. $(3.417 - 8.56) - [(-3.412) - (-8.807)]$.
7. $[(-1.314) + (-4.35)] + [(-3.89) - (2.003)]$.
8. $[(35 - 20) - 15] - [36 - (-23)]$.
9. (a) $(-\frac{8}{9}) + (-\frac{5}{7})$.   (b) $(\frac{4}{15}) - (-\frac{11}{12})$.
10. (a) $(-\frac{3}{8}) - (-\frac{23}{36})$.   (b) $(-\frac{7}{8}) - (\frac{21}{32})$.

## 10.2 Multiplication and Division

In this section, we use the properties of the positive rationals to motivate the definitions of multiplication and division on the set of all rational numbers. For example, let us consider what the product $(-4)(3)$ should be. To preserve the commutative property of multiplication, we want $(-4)(3)$ to be equal to $(3)(-4)$. With the interpretation of multiplication for *positive integers* as repeated addition, $(3)(-4)$ should be the sum $(-4) + (-4) + (-4) = -12$. Thus, we define $(-4)(3)$ to be $-12$.

Another way to motivate the definition for the product $(3)(-4)$ is to use the fact that we wish to preserve the distributive property of addition. From this point of view, the following shows that the preceding definition is consistent with the distributive property. Since $4 + (-4) = 0$ it follows that $3[4 + (-4)] = 3 \times 0$; that is, $3[4 + (-4)] = 0$. Using the distributive property, we obtain

$$3[4 + (-4)] = (3)(4) + (3)(-4).$$

Therefore,

$$(3)(4) + (3)(-4) = 0$$

or

$$12 + (3)(-4) = 0.$$

Consequently, $(3)(-4)$ is the additive inverse of 12; that is, $(3)(-4) = -12$.

We use a similar approach to determine how to define the product $(-3)(-4)$. If the product of any number and zero is to be zero, then

## 10.2 Multiplication and Division

$$(-4)[3 + (-3)] = (-4)(0),$$
$$(-4)(3) + (-4)(-3) = 0,$$

and

$$(3)(-4) + (-3)(-4) = 0.$$

Since $(3)(-4) = -12$, $(-3)(-4)$ is the additive inverse of $-12$; thus, $(-3)(-4) = 12$.

The general rules for multiplying rationals are as follows.

*Let a and b be positive rational numbers.*

**Rule 11.** $(-a)(0) = (0)(-a) = 0.$

**Rule 12.** $(a)(-b) = (-b)(a) = -(ab).$

**Rule 13.** $(-a)(-b) = (-b)(-a) = ab.$

### Examples

1. $(-7) \times 0 = 0.$
2. $(7)(-3) = -(7 \times 3) = -21.$
3. $(-6)(-3) = 6 \times 3 = 18.$
4. $(-6)(4) + (-2)(-5) = (-24) + 10 = -14.$

Notice that $(21) \div (-3) = (-21) \div (3) = -(21 \div 3)$ and $(-21) \div (-3) = (21) \div (3)$. Therefore,

$$(21) \div (-3) = (-21) \div (3) = -(21 \div 3) = -[(-21) \div (-3)].$$

Since $a \div b = \frac{a}{b}$, we conclude that

$$\frac{21}{-3} = \frac{-21}{3} = -\left(\frac{21}{3}\right) = -\left(\frac{-21}{-3}\right).$$

In general, with any rational number written in fractional notation, we can associate three signs: one with the numerator, one with the denominator, and one with the fraction. Any two can be changed and the fraction still represents the same rational number.

### Examples

1. $(-8) \div (2) = -4.$
2. $(-18) \div (-9) = 2.$

3. $-\dfrac{2}{3} = \dfrac{-2}{3} = \dfrac{2}{-3} = -\left(\dfrac{-2}{-3}\right).$

4. $\left(-\dfrac{4}{7}\right) \div \left(-\dfrac{2}{3}\right) = \left(-\dfrac{4}{7}\right) \times \left(-\dfrac{3}{2}\right) = \left(\dfrac{4}{7} \times \dfrac{3}{2}\right) = \dfrac{6}{7}.$

5. $\left(\dfrac{8}{15}\right) \div \left(-\dfrac{4}{5}\right) = \left(\dfrac{8}{15}\right) \times \left(-\dfrac{5}{4}\right) = -\left(\dfrac{8}{15} \times \dfrac{5}{4}\right) = -\dfrac{2}{3}.$

## Exercises I

In Exercises 1 through 10, perform the indicated operations.

1. (a) $(-18) \times (3)$. (b) $(23) \times (-5)$.
2. (a) $(-8) \times (-5)$. (b) $(27) \div (-9)$.
3. (a) $(-24) \div (6)$. (b) $(-36) \div (-12)$.
4. (a) $(-\tfrac{2}{3}) \times (\tfrac{3}{8})$. (b) $(-\tfrac{4}{9}) \times (-\tfrac{3}{5})$.
5. (a) $(\tfrac{2}{5}) \times (-\tfrac{15}{16})$. (b) $(-\tfrac{4}{11}) \div (\tfrac{8}{23})$.
6. (a) $(-\tfrac{4}{7}) \div (-\tfrac{8}{9})$. (b) $0 \div (-\tfrac{2}{3})$.
7. (a) $(\tfrac{-2}{-3}) \times (\tfrac{-4}{5})$, (b) $(\tfrac{-8}{15}) \times (\tfrac{6}{-7})$.
8. (a) $(\tfrac{-9}{-11}) \times (\tfrac{7}{6})$. (b) $(\tfrac{-2}{3}) \div (\tfrac{-3}{2})$.
9. (a) $(-\tfrac{2}{5})[(-\tfrac{3}{8}) + (\tfrac{1}{2})]$. (b) $(-\tfrac{4}{9})[(\tfrac{3}{5}) + (-\tfrac{7}{8})]$.
10. $[(-\tfrac{1}{2})(\tfrac{4}{5})] \div [(\tfrac{-1}{3}) + (\tfrac{2}{-5})].$

## Exercises II

In Exercises 1 through 4, show that $a(b + c) = ab + ac$.

1. $a = -3, b = -7, c = 8$.
2. $a = -9, b = -6, c = -3$.
3. $a = -8, b = 4, c = -11$.
4. $a = 6, b = -4, c = -3$.

In Exercises 5 through 8, perform the indicated operations.

5. $[(-\tfrac{2}{3}) + (\tfrac{7}{8})] \div [(\tfrac{5}{11}) + (-\tfrac{1}{2})].$
6. $[(-\tfrac{4}{9}) + (-\tfrac{3}{8})] \times [(-\tfrac{3}{4}) + (\tfrac{1}{3})].$
7. $(2.4)(3.7) + (-4.5)(-6.1).$

In Exercises 8 through 10, perform the indicated operation and give the best three-place decimal approximation of the answer.

8. $[(-3.9)(7.2)] \div [(-4.1)(-5.4)].$
9. $[(-4\tfrac{1}{2}) + (2\tfrac{1}{3})] \div (5\tfrac{5}{6}).$
10. $(-\tfrac{2}{3})[\tfrac{4}{9} + \tfrac{5}{6}] + \tfrac{3}{2}.$

## 10.3 Inequalities

Since we have extended the number system to include negative rationals it is important that we reconsider the properties of inequality. For *any* two rational numbers $a$ and $b$, $a < b$ if and only if there is a *positive* number $x$ such that $a + x = b$. This definition is consistent with the one previously given and with the geometric interpretation that $a < b$ if and only if $a$ is the left of $b$ on the number line.

The set of all rational numbers possesses the important *trichotomy property* ($a < b$, $a = b$, or $a > b$) and the *transitive property* ($a < b$ and $b < c$ implies $a < c$) of inequality. Furthermore, it should be obvious that the additive properties of inequality ($a < b$ implies $a + c < b + c$, and $a < b$ and $c < d$ implies $a + c < b + d$) can be proved exactly as before. However, the multiplicative property that if $a < b$ then $ac < bc$ is not generally true. If $a < b$ and $c > 0$ then $ac < bc$; this is proved as before. If $a < b$ and $c < 0$, then $ac > bc$; this is proved as follows.

**Theorem.** If $a < b$ and $c < 0$, then $ac > bc$.
  *Proof.* If $c < 0$, then $(-c) + c < (-c) + 0$ and $0 < (-c)$. Since $a < b$ and $(-c) > 0$, it follows that $a(-c) < b(-c)$. Thus, $-(ac) < -(bc)$. Adding $ac$ and $bc$ to both sides of the last inequality, we obtain $bc < ac$, or $ac > bc$.

For the set of rationals, it is not generally true that $\frac{a}{b} < \frac{c}{d}$ if and only if $ad < bc$. However, if $bd > 0$ (that is, $b$ and $d$ have the same sign), then $\frac{a}{b} < \frac{c}{d}$ if and only if $ad < bc$. This is proved by multiplying both sides of $\frac{a}{b} < \frac{c}{d}$ by $bd$ and by dividing both sides of $ad < bc$ by $bd$.

Basically, the new fact we need to remember about inequalities is that when multiplying or dividing both sides of an inequality by a negative number it reverses the sense of the inequality.

### Exercises I

In Exercises 1 through 10, state whether or not each statement is true.

1. $-3 < 0$.
2. $-27 < -2$.

3. $-1000 > 31$.
4. $-\frac{2}{3} < \frac{-5}{8}$.
5. $-\frac{11}{37} < -\frac{1}{9}$.
6. $-\frac{5}{7} < -\frac{3}{5}$.
7. $\frac{-3}{11} < \frac{-5}{12}$.
8. $\frac{1}{8} - \frac{5}{6} < \frac{2}{7} - \frac{2}{3}$.
9. $(-\frac{3}{8})(\frac{5}{7}) < (-\frac{3}{8})(\frac{6}{7})$.
10. $\frac{2}{9} - \frac{4}{7} < \frac{5}{12} - \frac{8}{15}$.

### Exercises II

1. Is $\frac{23}{36} - \frac{5}{11} < \frac{5}{12} - \frac{9}{22}$?
2. Is $(-\frac{2}{5}) + (-\frac{7}{12}) < (-\frac{8}{9}) + (\frac{1}{6})$?
3. Is $(-\frac{4}{5}) - (-\frac{2}{3}) < (\frac{5}{7}) - (\frac{1}{14})$?

Prove each of the theorems in Exercises 4 through 10 where $a$, $b$, and $c$ are rational numbers.

4. If $a < b$ then $a + c < b + c$.
5. If $a < b$ and $c < d$, then $a + c < b + d$.
6. If $a < b$ and $c > 0$, then $ac < bc$.
7. If $ad < bc$ and $bd > 0$, then $\frac{a}{b} < \frac{c}{d}$.
8. If $\frac{a}{b} < \frac{c}{d}$ and $bd > 0$, then $ad < bc$.
9. If $a < b$ and $ab > 0$ then $\frac{1}{a} > \frac{1}{b}$.

## 10.4 Supplementary Exercises

1. Add: (a) $-58$ / $42$    (b) $-83$ / $97$    (c) $-34$ / $-51$
2. Add: (a) $29$ / $-37$    (b) $-56$ / $-35$    (c) $68$ / $-59$
3. Subtract: (a) $-89$ / $37$    (b) $-63$ / $-43$    (c) $-57$ / $-78$
4. Subtract: (a) $-37$ / $-51$    (b) $85$ / $98$    (c) $-37$ / $-18$

In Exercises 5 through 10, perform the indicated operations.

5. $5.723 - 8.413$.
6. $(-4.23) - (-5.67)$.
7. $8.417 - (-3.568)$.
8. $(-8.71) - (8.38)$.
9. $87.413 - 97.430$.
10. $(-23.67) + (63.79)$.

# 11

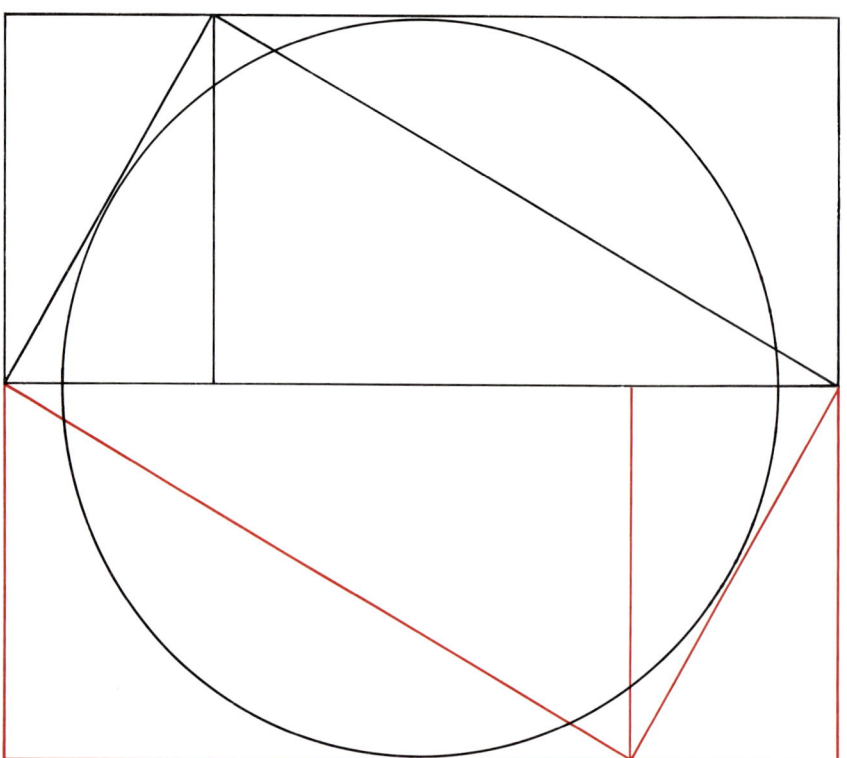

# APPLICATIONS

## 11.1 Introduction

The usefulness of numbers in the practical problems of man's everyday pursuits was the primary motivation for the development of the number systems. The uses we make of numbers are so numerous and varied that it would be nearly impossible to list them all. Numbers are important in statistical work such as that done by insurance companies; they are indispensible in commerce and business transactions; and they are vital as measures of length, area, and volume.

In this chapter, we shall discuss several of the important practical applications of numbers. The following list of questions should give the student an idea of the kinds of problems we intend to discuss.

1. If a $10\frac{3}{4}$ ounce package of a given product is selling for 42 cents and the $5\frac{5}{8}$ ounce package of the same product is selling for 22 cents, which is the better buy?
2. (a) If the grades made on the first five tests a student takes are 68, 97, 84, 78, and 93, what is his grade average? (b) If the student takes another test and if all six tests are weighted equally, is it possible for him to raise his average by 3 points if 100 is the highest grade that can be made on the sixth test?
3. If a runner can run 100 yards in 10 seconds, what is his average speed in miles per hour?
4. Suppose a room is 9 feet wide and 11 feet 3 inches long. If the floor is to be tiled with tiles which are 9 inches square and which cost 22 cents each, what is the cost of the tile needed to cover the floor?

5. How long would it take to double an amount of money put on deposit in a savings account if the interest were compounded quarterly at $5\frac{1}{4}$ percent?
6. If a person saves five cents on a given day, ten cents the next day, fifteen cents the third, twenty cents the fourth, etc., what is the amount saved in thirty days if the savings program is maintained?
7. Which would be the better one-year investment: $1,000 drawing 5 percent compounded quarterly or $1,000 drawing $5\frac{1}{4}$ percent simple interest? How much better?
8. Two trucks must make deliveries totaling 100 miles. Truck A gets 15 miles per gallon of fuel and truck B gets 10 miles per gallon. If the total fuel consumption is to be no more than 8 gallons, what is the minimum number of miles truck A must travel?
9. How many degrees are in the acute angle formed by the hands of a clock when it is 10 minutes after 4 o'clock?
10. Suppose the total number of units that can be produced in a given day by Mr. Brown and Mr. Smith working at separate machines is 100. If it is known that Mr. Brown working 5 days and Mr. Smith working 3 days can produce no more than 424 units of the product, what is the least number of units that could be produced by Mr. Brown in one day? (We assume each man produces the same number each day.)

## 11.2 Linear Measure

In order to measure distance, it was necessary for man to establish a unit of linear measure. The choice one makes for a unit of linear measure is not a natural one in the sense that the day is a natural unit of measure for time. In fact, until science and technology advanced to a point where precision tools and accurate measurements were essential, a method to perpetuate the exactness of the unit chosen was not of such vital importance.

In this country we use two systems of linear measure: the *English system* and the *metric system*. As we shall see, the metric system is much better suited to our base-ten number system. However, even though the world has chosen the meter as the principal unit of length, the English system is still an important system of measure which is used extensively in this country.

## 11.2 Linear Measure

The *yard* is a unit of linear measure in the English system that was standardized by Henry I of England; it was supposed to be the distance from his nose to the tips of his outstretched fingers. The unit called the *foot* has obvious origins. The unit known as the *mile* dates back to the Romans; it originated with *mille passum* which meant a thousand paces. A Roman "pace" was roughly five feet (two steps) so *mille passum* (or a mile) was roughly five thousand feet.

Today, our units have been officially established by Congress in terms of the meter. An *inch* is a length such that 39.37 inches is exactly one meter in length. We shall discuss below how the meter was selected and how its exactness is maintained.) Once the inch was established, other units were established in terms of this chosen unit. The units in the English system are given as follows.

**English System**
1 foot = 12 inches
1 yard = 3 feet
1 rod = 16.5 feet
1 mile = 5,280 feet

Since 12 inches = 1 foot, 3 inches is equivalent to $\frac{3}{12}$ or $\frac{1}{4}$ of a foot. Similarly, $\frac{1}{3}$ foot = 4 inches, $\frac{1}{2}$ foot = 6 inches, and $\frac{1}{2}$ yard = $1\frac{1}{2}$ feet. Since 3 feet = 1 yard, there are $5,280 \div 3 = 1,760$ yards in a mile.

Although most of us are familiar with these units in the English system, we still have difficulty comprehending some of the measurements that modern-day scientists can make with these units. For example, astronomers have determined that one of the stars, Alpha Centauri, nearest to our earth is twenty-four trillion miles away. Although conceptually it may not help, scientists have created another unit called a *light year* with which to express many such celestial measurements. A light year is the distance that light travels in one year; a light year is about six trillion miles. Thus, Alpha Centauri is about four light years from the earth, an almost inconceivable distance.

As stated earlier, the inch was established in terms of the meter. Originally, the meter was defined as one ten-millionth of the distance from the North Pole to the Equator. This unit was later standardized and was defined as the distance between specified markings on a platinum bar deposited at the International Bureau of Weights and Measures in Paris. The bar is kept at 0° centigrade in order to ensure its constant length and the exactness of the meter.

Actually, the man-made bar chosen to maintain the exactness of the meter is no longer necessary since the meter has been shown to be equivalent to the length of 1,552,734.8 waves of red cadmium light, a length whose variance is not in question.

The meter is divided into *one hundred* equal lengths called *centimeters*. Again, each centimeter is divided into *ten* equal lengths which are called *millimeters*; a millimeter is therefore one-thousandth of a meter. The units in the metric system are given as follows.

**Metric System**

10 millimeters   = 1 centimeter
10 centimeters   = 1 decimeter
10 decimeters    = 1 meter
100 centimeters  = 1 meter
1,000 meters     = 1 kilometer

## *Examples*

1. How many inches in a mile? *Solution.* Since

$$1 \text{ ft} = 12 \text{ in.,}$$

it follows that

$$5{,}280 \times 1 \text{ ft} = 5{,}280 \times 12 \text{ in.,}$$
$$1 \text{ mile} = 63{,}360 \text{ in.}$$

2. How many miles is 7,200 feet? *Solution.* Since

$$5{,}280 \text{ ft} = 1 \text{ mile,}$$

it follows that

$$1 \text{ ft} = \frac{1}{5{,}280} \text{ mi.}$$

Thus,

$$7{,}200 \times 1 \text{ ft} = 7{,}200 \times \frac{1}{5{,}280} \text{ mi,}$$

$$7{,}200 \text{ ft} = \frac{720}{528} \text{ mi,}$$

$$7{,}200 \text{ ft} = \frac{15}{11} \text{ mi} = 1\frac{4}{11} \text{ mi.}$$

## 11.2 Linear Measure

3. How many feet in a kilometer? *Solution.*

$$39.37 \text{ in.} = 1 \text{ meter},$$
$$1{,}000 \times 39.37 \text{ in.} = 1{,}000 \times 1 \text{ meter},$$
$$39{,}370.00 \text{ in.} = 1{,}000 \text{ meters},$$
$$39{,}370 \text{ in.} = 1 \text{ kilometer}.$$

Since

$$1 \text{ in.} = \frac{1}{12} \text{ ft},$$

$$39{,}370 \text{ in.} = \frac{39{,}370}{12} \text{ ft};$$

thus,

$$39{,}370 \text{ in.} = 3{,}280\tfrac{5}{6} \text{ ft},$$

and

$$1 \text{ kilometer} = 3{,}280\tfrac{5}{6} \text{ ft}.$$

We often write $12\tfrac{\text{in.}}{\text{ft}}$, or 12 in./ft (read "twelve inches per foot"), or $3\tfrac{\text{ft}}{\text{yd}}$ (read "three feet per yard") to indicate the relationship between the given units. This notation has some arithmetic advantages in working problems provided we treat, for example, $\tfrac{\text{ft}}{\text{yd}}$ as a fraction. Consider the following example. Since there are $5{,}280\tfrac{\text{ft}}{\text{mi}}$, we conclude that there are

$$5{,}280 \frac{\text{ft}}{\text{mi}} \times 12 \frac{\text{in.}}{\text{ft}} = 5{,}280 \times 12 \times \frac{\text{ft}}{\text{mi}} \times \frac{\text{in.}}{\text{ft}} = 63{,}360 \frac{\text{in.}}{\text{mi}}.$$

### Exercise I

1. (a) Which is the longer distance, 100 yards or 100 meters?
   (b) How many inches difference are there between 100 yards and 100 meters?
2. Is $\tfrac{5}{8}$ of a mile longer or shorter than a kilometer?
3. (a) In meters, what is the approximate distance from the North Pole to the Equator? (Use the original definition of the meter.)
   (b) Assuming the distance in part (a) is one-fourth of the circumference of the earth (distance around), what is the circumference of the earth in meters?

(c) Using your answer in part (b), find the circumference in miles.
4. How many centimeters are there in an inch?
5. (a) Express 15 meters in terms of centimeters.
   (b) Express 15 feet in terms of inches.
6. (a) Express 23 kilometers in terms of meters.
   (b) Express 23 miles in terms of feet.
7. (a) Express 537 centimeters in terms of meters.
   (b) Express 537 inches in terms of feet.
8. (a) Express 23,471 meters in terms of kilometers.
   (b) Express 23,471 feet in terms of miles.
9. (a) How many rods in one mile?
   (b) How many rods in 38 yards?
10. Which is longer, $\frac{1}{16}$ inch or one millimeter?

## 11.3 Area

A *polygon* is a many-sided closed figure made up of line segments. Several examples of polygons are given in Figure 1. Obviously, for the figure to be closed, any polygon has at least three sides; thus, the triangle is the polygon with the least number of sides.

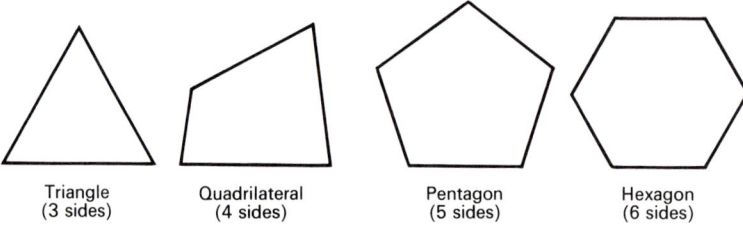

Triangle (3 sides)  Quadrilateral (4 sides)  Pentagon (5 sides)  Hexagon (6 sides)

Figure 1

A polygon is said to be *regular* if all sides are of equal length and all interior angles are equal. For example, the polygon in Figure 2 is a regular octagon.

Certain quadrilaterals (polygons with four sides) are important enough to deserve special names. A quadrilateral with each of the four interior angles a right angle is called a *rectangle* A rectangle with all four sides equal is called a *square*. Any quadrilateral with

## 11.3 Area

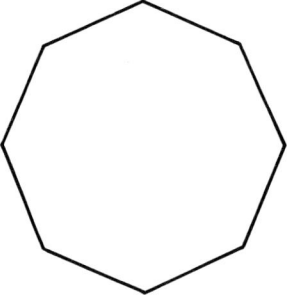

Figure 2. *Regular octagon (eight sides of equal length).*

two pairs of parallel sides is called a *parallelogram* and a quadrilateral with exactly one pair of parallel sides is called a *trapezoid*. We should note that these terms are not exclusive since a square is a rectangle and both rectangles and squares are parallelograms. From our definition of trapezoid, a square, for example, would not be a trapezoid; however, if we had said that a trapezoid was a quadrilateral with at least one pair of parallel sides, then a square, a rectangle, and a parallelogram would each be a trapezoid. Generally, the parallel sides of a trapezoid are called the *bases*. (See Figure 3.)

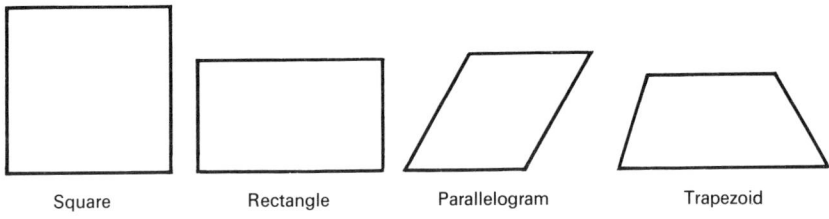

Figure 3

As we said in Chapter 7, the region enclosed in a square with each side of unit length is our unit of measure for area. We concluded that a rectangle with length $l$ and width $w$ had an area of $lw$ square units. For example, a rectangle with length 5 inches and width 4 inches has an area of 20 square inches.

Figure 4 should clearly indicate how the formulas for the areas of parallelograms, trapezoids, and triangles are obtained.

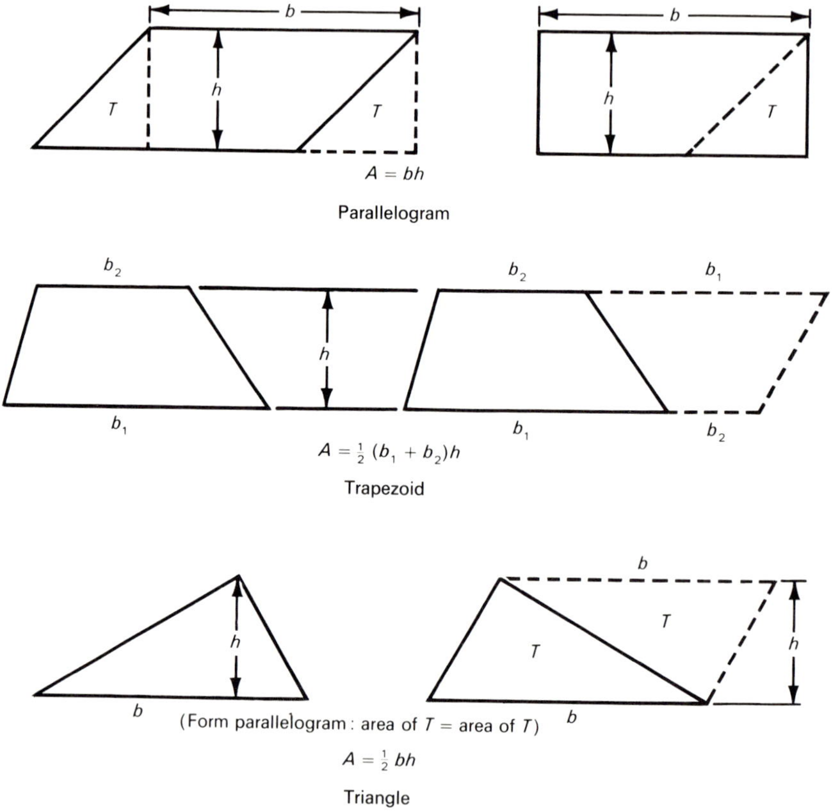

b
h
T
T
A = bh
Parallelogram
b
h
T
b₂
h
b₁
b₂
b₁
b₁
b₂
A = ½ (b₁ + b₂)h
Trapezoid
h
b
b
T
T
h
b
(Form parallelogram: area of T = area of T)
A = ½ bh
Triangle

Figure 4

One can derive formulas for areas of other polygonal regions, but often it is sufficient if we realize that every such region can be subdivided into triangular regions. Therefore, the area of a region bounded by a polygon can often be obtained by finding the sum of the areas of the triangular regions formed. (See Figure 5.)

A₁
A₂
A₃
A₄

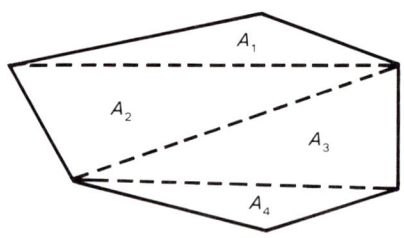

Figure 5. $A = A_1 + A_2 + A_3 + A_4$.

## 11.3 Area

Since a *square foot* is a unit of area equivalent to a square region with each side one foot in length and since there are 12 inches in a foot, we conclude that there are 12 × 12 square inches in a square foot. Other units of measure for area are the square yard, square rod, and square mile. (Of course, there are similar units for the metric system.) Another common unit of measure for area is called the *acre*; an acre is a unit of land measure such that 640 acres is equivalent to one square mile. Following is a list of some of the equivalent measures for area in the English system.

**Equivalent Measures for Area**
144 sq in. = 1 sq ft
9 sq ft = 1 sq yd
30.25 sq yd = 1 sq rod
640 acres = 1 sq mile

*Examples*

What is the area of a rectangular region whose length is 7 feet 4 inches and width is 4 feet 9 inches? *Solution.* If we desire to obtain the area in square inches, we can change the linear measurements to inches. Since

$$7 \text{ ft } 4 \text{ in.} = (7 \times 12) + 4$$
$$= 84 + 4$$
$$= 88 \text{ in.}$$

and

$$4 \text{ ft } 9 \text{ in.} = (4 \times 12) + 9$$
$$= 48 + 9$$
$$= 57 \text{ in.},$$

the area is 88 × 57 = 5,016 sq in.

Of course, we could change each dimension to feet. Since

$$7 \text{ ft } 4 \text{ in.} = 7\frac{4}{12}$$
$$= 7\frac{1}{3} \text{ ft}$$

and

$$4 \text{ ft } 9 \text{ in.} = 4\frac{9}{12}$$
$$= 4\frac{3}{4} \text{ ft,}$$

the area is

$$7\frac{1}{3} \times 4\frac{3}{4} = \frac{22}{3} \times \frac{19}{4}$$
$$= \frac{418}{12}$$
$$= 34\frac{5}{6} \text{ sq ft.}$$

Since there are 144 sq in. in 1 sq yd,

$$34\frac{5}{6} \text{ sq ft} = 34\frac{5}{6} \times 144$$
$$= \frac{209}{6} \times 144$$
$$= 209 \times 24$$
$$= 5{,}016 \text{ sq in.}$$

Thus, we have obtained the area in square inches by two methods; to do this is one way to check the accuracy of answers.

A plane figure other than a polygon which has important applications is the circle. A *circle* is the collection of all points in a plane which are equidistant from a fixed point within called the *center*. The distance from the center of the circle to the circle is called the *radius*. Unfortunately, since a circular region cannot be partitioned into triangular regions as can a polygonal region, the derivation of the formula for the area of a circular region is quite difficult. In fact, even though the radius is a rational length such as 3 inches, the number of square inches in the area of a circular region can be shown not to be a rational number. (We shall have more to say about such numbers in Section 12.8.) The area of a circular region is the product of the radius squared and a number which we call *pi*, denoted by $\pi$. A four-place decimal approximation of $\pi$ is 3.1416. The *circumference* of a circle is the distance around the circle and a formula to find the circumference also is given in terms of $\pi$; the circumference is $2\pi r$ where $r$ is the radius.

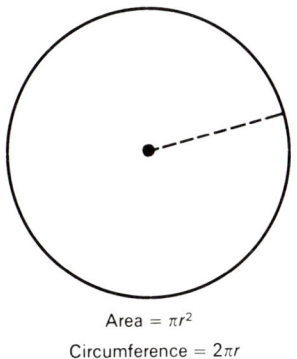

Area = $\pi r^2$
Circumference = $2\pi r$

## Exercises I

1. How many square feet are in an acre?
2. How many square yards are in an acre?
3. How many square rods are in an acre?

Find the area of each of the following regions bounded by polygons with the given dimensions in Exercises 4 through 9.

4. A triangle with a 16 inch base and a height of $17\frac{3}{4}$ inches.
5. A rectangle whose length and width are $25\frac{1}{2}$ inches and $11\frac{1}{4}$ inches, respectively.
6. A trapezoid whose bases are $18\frac{1}{2}$ inches and $12\frac{2}{3}$ inches and whose height is 8 inches.
7. A triangle with a 1 ft 2 in. base and a height of $9\frac{5}{6}$ in.
8. A rectangle whose length and width are $5\frac{3}{4}$ ft and 2 yd $1\frac{1}{2}$ ft, respectively.
9. A trapezoid whose bases are 2 ft 5 in. and $42\frac{2}{3}$ in. and whose height is 38 in.
10. A room is 23 ft 3 in. long and 11 ft 9 in. wide.
    (a) Find the area of the floor in square inches.
    (b) Give the area in square feet.
    (c) Give the area in square yards.
11. A room is 20 ft 6 in. long and 12 ft 3 in. wide.
    (a) Find the area of the floor in square inches.
    (b) Give the area in square feet.
    (c) Give the area in square yards.

12. The area of a rectangular region is 238 sq in. If the length is $18\frac{1}{2}$ in., find the width.
13. The area of a rectangular region is 124 sq ft. If the length is $4\frac{2}{3}$ yd, find the width.
14. Find the area and circumference of a circle with a radius of 23 centimeters.
15. (a) Find the area in square inches of a circular region with a radius of 1 ft 2 in.
    (b) Find the circumference in inches of a circular region with a radius of 2 ft 2 in.
16. If the circumference of a circle is $32\pi$ inches, what is the radius?

If the perimeter of a polygon (the distance around) is the total length of all the sides, find the perimeter of each of the polygons in Exercises 17 through 19.

17. A rectangle which is 3 ft 2 in. in length and 5 ft 7 in. in width.
18. A square which is 5.37 cm on a side.
19. A regular octagon which is 1 ft 8 in. on a side.
20. The perimeter of a rectangle is 391 inches. If the width is 63 inches, find the length.
21. The perimeter of a rectangle is 412 inches. If the width is 2 ft 7 in., find the length.
22. Suppose a field is 128 ft wide and 220 ft long.
    (a) Find the amount of fence needed to enclose the field.
    (b) What is the area of the field in square feet?
    (c) What is the area of the field in square yards?
    (d) What is the area of the field in square rods?
    (e) How many acres are there in the field?
23. A 120 acre farm which is rectangular in shape fronts on a straight road for $\frac{1}{4}$ mile. How many yards does the farm extend back from the road?

## 11.4 Volume

As *area* is a measure for an enclosed region in two dimensions, *volume* is a measure of an enclosed region in three dimensions. Similar to using a unit square to measure area, we use a unit cube to measure volume. The *unit cube* is a six-sided solid called a *polyhedron* with each face being a unit square.

## 11.4 Volume

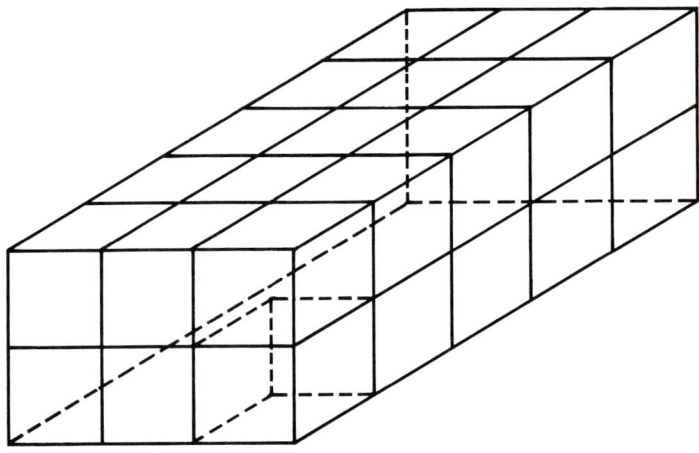

Figure 6. $lwh = 5 \times 3 \times 2 = 30$ *cubic units*.

It should be obvious by observing Figure 6 that the number of these unit cubes contained in a rectangular box which is *l* units long, *w* units wide, and *h* units high is *lwh*. In general, the volume *V* of a rectangular box with length *l*, width *w*, and height *h* is given by

$$V = l \times w \times h.$$

Since a rectangular box has two faces with area $h \times w$, two faces with area $l \times w$, and two with area $l \times h$, the *total surface area S* is given by

$$S = 2(hw + lw + lh).$$

### Examples

1. A room is 9 ft wide, 15 ft long, and 8 ft high. (a) Find the number of cubic feet in the room. (b) Find the number of square feet in the walls and ceiling.
   Solution. (a) $9 \times 15 \times 8 = 1{,}080$ cu ft. (b) The area contained in the four walls and ceiling is $2hl + 2hw + wl$; that is,
   
   $$2(8)(15) + 2(8)(9) + (9)(15) = 240 + 144 + 135 = 519 \text{ sq ft.}$$

2. If water weighs 62.5 lb per cubic foot, how much does the water in a swimming pool weigh if it is 20 ft wide, 45 ft long, and has a uniform depth of 6 ft?

*Solution.* The volume of water is 6 × 20 × 45 = 5400 cu ft. Since each cubic foot weighs 62.5 lb, the weight of the water is 62.5 × 5400 = 337,500 lb.

A *sphere* is the collection of all points in three dimensions which are equidistant from a point within called the *center* of the sphere. The distance from the center to the surface of the sphere is called the *radius*. Both the surface area and the volume of a sphere are given in terms of the number $\pi$ which was used to determine the area and circumference of a circle. If $r$ is the radius of a sphere, $S$ the surface area, and $V$ the volume (Figure 7), then

$$V = \frac{4}{3}\pi r^3 \quad \text{and} \quad S = 4\pi r^2.$$

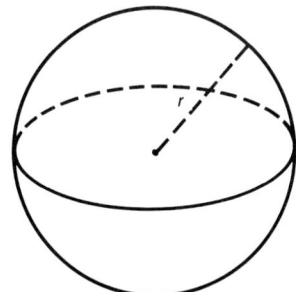

Figure 7. *Sphere.*

## Examples

1. Assume $\pi = 3.14$ and that the radius of the earth is 4,000 miles. Find the surface area of the earth.
   *Solution.* The surface area is $4(3.14)(4,000)^2 = 200,960,000$ sq mi.
2. Find the volume of the earth.
   *Solution.* The volume is $\frac{4}{3}(3.14)(4,000)^3 = 267,946,666,666$ cu mi approximately; that is, it is approximately 270,000,000,000 or $2.7 \times 10^{11}$ cu mi.

Another solid which we encounter quite often is a right circular cylinder (Figure 8); a tin can has the shape of a right circular cylinder. The volume $V$ of a right circular cylinder is given by

$$V = \pi r^2 h,$$

where $r$ is the radius and $h$ is the height.

## 11.4 Volume

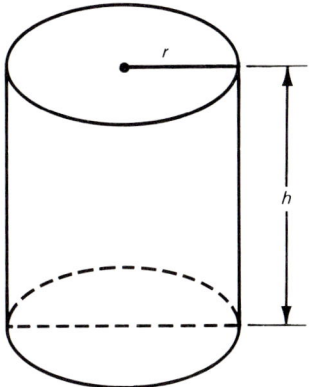

Figure 8. *Right circular cylinder.*

The formula for the surface area of a tin can is not difficult to derive. Since both the top and bottom are circular regions, the area of each is $\pi r^2$; thus, the total surface area of the top and bottom is $2\pi r^2$. If we remove the top and bottom, make one vertical cut in the can, and flatten it out, we have a rectangular region whose length is the circumference of the circular base and whose width is the height of the can. Therefore, the area which is called the *lateral surface area* of the cylinder is $2\pi rh$. The total surface area of the can is given by

$$S = 2\pi r^2 + 2\pi rh.$$

## Example

A coffee can is $5\frac{1}{2}$ in. high and has a 2 in. radius. (a) What is the total surface area? (Let $\pi = 3.14$.) (b) What is the volume? *Solution.* The surface area is $2\pi r^2 + 2\pi rh = 2(3.14)(2)^2 + 2(3.14)(2)(\frac{11}{2}) = 25.12 + 69.08 = 94.20$ sq in., approximately. (b) The volume of the can is approximately $\pi r^2 h = (3.14)(2)^2(\frac{11}{2}) = 148.16$ cu in.

### Exercises I

In Exercises 1 through 3, find the volume and surface area of each rectangular box with the given dimensions.
  1. $l = 3\frac{1}{2}$ ft, $w = 2\frac{1}{3}$ ft, $h = 6$ ft.

2. $l = 1$ ft 4 in., $w = 2$ ft 8 in., $h = 9$ in.
3. $l = 2$ ft 5 in., $w = 3\frac{1}{6}$ ft, and $h = 11$ in.

In Exercises 4 through 6, find the volume and surface area of each right circular cylinder with the given radius and height. (Use $\pi = 3.14$.)

4. $r = 8$ in. and $h = 1\frac{1}{4}$ ft.
5. $r = 2$ ft 3 in. and $h = 3$ ft 5 in.
6. $r = 22$ in. and $h = 2$ ft 9 in.

In Exercises 7 through 9, find the volume and surface area of each sphere with the given radius. (Use $\pi = 3.14$.)

7. $r = 4$ in.   8. $r = 1$ ft 2 in.   9. $r = 7\frac{5}{8}$ in.
10. Find the volume of air in a room which is 9 ft wide, 11 ft 3 in. long, and 9 ft high.
11. For the room in Exercise 10, what is the total surface area?
12. A rectangular room with floor dimensions of 9 ft by 12 ft has a volume of 918 cu ft. What is the height of the room?
13. About how many cubic centimeters are in a cubic inch?
14. A cubic foot of water weighs 62.5 lb. If lead is approximately 11.4 times as heavy as water, what does a cubic yard of lead weigh?
15. Find the surface area of the moon assuming that the radius of the moon is 1,080 miles.
16. Find the volume of the moon assuming that the radius is 1,080 miles.
17. If the edges of each face of a cube are doubled in length, how many times is the volume increased?
18. If the radius of a sphere is doubled, how many times is the volume increased?
19. A cylindrical pipe open at both ends is 8 ft long. If its inner radius is $3\frac{1}{2}$ in. and its outer radius is 4 in., find the volume of the metal needed to make the pipe.
20. A rectangular-shaped room is 12 ft wide and 16 ft long. If the total surface area is 1,154 sq ft, what is the height of the room?

## 11.5 Other Measures

One could devote an entire book to the study of various measures and units of measure used throughout the world. A table just to

## 11.5 Other Measures

list the interrelationship between all comparable units would require many pages. Thus, we shall discuss only some of the most important measures and the most familiar units. (Some units which we do not discuss but which the student might like to investigate on his own are the radian, calorie, horsepower, decibel, carat, and grain.)

Besides the measurement of angles, some other types of measure which we have not already discussed pertain to *capacity*, *weight*, and *time*. The standard units for measure of capacity are given in the following table.

| Liquid Measure | Dry Measure |
|---|---|
| 1 pint = 16 fluid ounces | 1 bushel = 4 pecks |
| 2 pints = 1 quart | 1 peck = 8 quarts |
| 4 quarts = 1 gallon | 1 quart = 2 pints |

Although "pint" and "quart" are used for both liquid and dry measure, we should point out that as units of measure they are quite different.

### Example

If a cylindrical tank has a radius of 2 ft and is 6 ft long, how many quarts of oil will it hold if $7\frac{1}{2}$ gallons fill one cubic foot?
*Solution.* The volume of the tank is $\pi r^2 h = 24\pi$ cubic feet. Since one cubic foot holds $7\frac{1}{2}$ gallons, $24\pi$ cubic feet hold $24\pi \times \frac{15}{2} = 12\pi \times 15 = 180\pi$ gallons. Since there are 4 quarts in one gallon, the tank holds $4 \times 180\pi = 720\pi$ quarts; that is, approximately $720 \times 3.14 = 2{,}256.8$ quarts.

The standard units of weight are given in the following table.

| Weights |
|---|
| 1 pound = 16 ounces |
| 1 ton = 2,000 pounds |

### Example

Which is the greater weight, $3\frac{5}{8}$ lb or 60 oz?

*Solution.* Since there are 16 ounces in a pound, $3\frac{5}{8}$ pounds is equivalent to

$$3\frac{5}{8} \times 16 = \frac{29}{8} \times 16 = 58 \text{ ounces.}$$

Thus, 60 oz. is the greater weight.

Although we have a conglomeration of units for weights throughout the world, we are fortunate that the units for time are standardized throughout the civilized world. The table for units of time is as follows.

*Time*

| |
|---|
| 60 seconds = 1 minute |
| 60 minutes = 1 hour |
| 24 hours = 1 day |
| 7 days = 1 week |

## *Example*

How many seconds are there in the month of January?

*Solution.* Since there are 60 seconds in a minute and 60 minutes in an hour, there are $60 \times 60 = 3600$ seconds in an hour. Since there are 24 hours in one day, there are $24 \times 360 = 86{,}400$ seconds in a day. Since there are 31 days in January, there are $31 \times 86{,}4000 = 2{,}678{,}400$ seconds in the month of January.

When a car travels 50 miles in one hour, we say the average velocity is 50 miles per hour (often written 50 mi/hr, or 50 $\frac{\text{mi}}{\text{hr}}$). Average velocity is defined to be the distance traveled divided by the time; that is,

$$\text{Average velocity} = \frac{\text{distance}}{\text{time}}.$$

If a car travels 72 miles in $1\frac{1}{2}$ hours, its average velocity is

$$72 \div \frac{3}{2} = 72 \times \frac{2}{3} = 48 \tfrac{\text{mi}}{\text{hr}}.$$

We often let $v$ represent velocity, $d$ represent distance, and $t$ represent time. In this case, we have

## 11.5 Other Measures

$$v = \frac{d}{t} \quad \text{or} \quad d = vt.$$

Consequently, we see that distance traversed is equal to the product of the rate (velocity) and the time.

Of course, one can give velocity in various units besides miles per hour; for example, velocity might be given in feet per second, miles per second, centimeters per second, etc. Generally, the choice is determined by the application.

### Examples

1. What velocity in feet per second is equivalent to 60 $\frac{mi}{hr}$?
   *Solution.* Since there are 5,280 feet in one mile, 60 miles is $60 \times 5,280 = 316,800$ ft. Since there are 3600 seconds in an hour,
   $$v = \frac{316,800}{3600} = 88 \tfrac{ft}{sec}.$$

2. If a man can run 100 yards in 10 seconds, what is his average velocity in miles per hour?
   *Solution.*
   $$v = \frac{100 \text{ yd}}{10 \text{ sec}} = 10 \tfrac{yd}{sec};$$
   thus,
   $$v = 10 \tfrac{yd}{sec} \times 3600 \tfrac{sec}{hr} = 36,000 \tfrac{yd}{hr}.$$
   Since there are 1,760 yards in one mile,
   $$v = 36,000 \tfrac{yd}{hr} \div 1,760 \tfrac{yd}{mi}$$
   $$= 36,000 \tfrac{yd}{hr} \times \frac{1}{1,760} \tfrac{mi}{yd}$$
   $$= \frac{36,000}{1,760} \tfrac{mi}{hr}$$
   $$= 20 \tfrac{5}{11} \tfrac{mi}{hr}.$$

Consider a circle with the circumference divided into 360 equal arcs. The measure of the angle subtended at the center by the ends of one such arc is called a *degree*. (See Figure 9.) Since $\frac{90}{360} = \frac{1}{4}$, we see that 90 degrees is the measure of the angle subtended at the

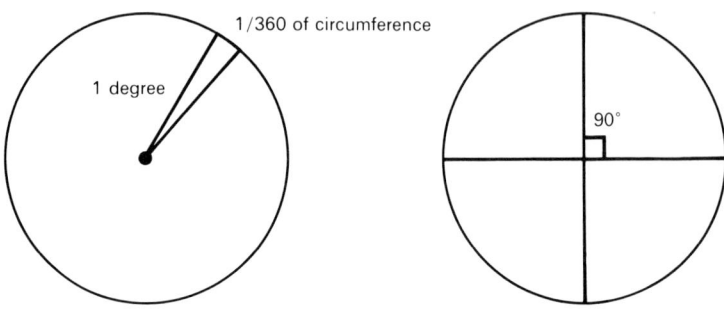

Figure 9. *Degrees.*

center of the circle by an arc which is $\frac{1}{4}$ the length of the circumference of the circle. An angle with a measure of 90 degrees, denoted by 90°, is called a *right angle*; the sides of a right angle are *perpendicular* to each other. An angle whose measure is less than a right angle is called an *acute angle*.

### Example

How many degrees in the acute angle formed by the hands of a clock when it is 10 minutes after 4 o'clock?

*Solution.* Since the circular face on the clock is divided into 12 equal parts (Figure 10), the angle with vertex at the center and sides through any two adjacent hour numbers has a measure of $\frac{360}{12} = 30$ degrees. Since 10 minutes is $\frac{10}{60} = \frac{1}{6}$ of an hour, the measure of the angle at ten after four with the hour hand as one side and a ray from the center through 4 as the other is $\frac{1}{6} \times 30 = 5$ degrees. The number of degrees in the angle between the minute hand which

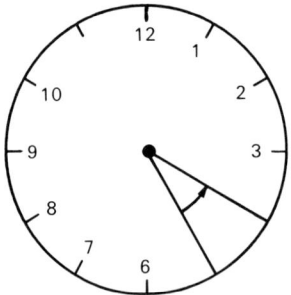 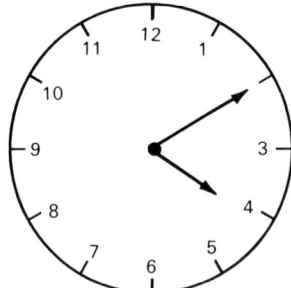

Figure 10

11.6 Miscellaneous Applications

points at 2 at 10 minutes after the hour and the segment connecting the center and the point at 4 on the face is $2 \times 30 = 60$. Thus, the number of degrees in the acute angle formed by the two hands is $60 + 5 = 65$ degrees.

### Exercises I

1. (a) How many quarts in $3\frac{1}{2}$ gallons?
   (b) How many pints in $3\frac{1}{2}$ gallons?
2. How many ounces in 12.5 pounds?
3. (a) How many minutes in 40 hours?
   (b) How many seconds in 40 hours?
4. What velocity in feet per second is equivalent to 50 $\frac{mi}{hr}$?
5. If a man can run one mile in 5 minutes, what is his average velocity in miles per hour?
6. (a) A person drives 10 miles at 40 $\frac{mi}{hr}$ and another 10 miles at 60 $\frac{mi}{hr}$. What is the total time required to drive the 20 miles?
   (b) At what constant velocity would a person need to drive in order to go the 20 miles in the same amount of time?
7. Through how many degrees does the hour hand on a clock turn in 3 hours and 18 minutes?
8. What is the measure of the smaller angle formed by the hands on a clock at 40 minutes after 11 o'clock?
9. On a 40 mile drive, how many minutes would be saved by driving 75 $\frac{mi}{hr}$ instead of 70 $\frac{mi}{hr}$?
*10. If the radius of a tire is 16 inches, approximately how many complete revolutions will the wheel make in one minute when the car is traveling 60 $\frac{mi}{hr}$?

## 11.6 Miscellaneous Applications

A little use of arithmetic often pays dividends in daily buying. For example, one supermarket is selling olives in a 5 ounce jar at 35 cents. The "large economy" size is a 9 ounce jar selling at 65 cents. There is no question that the second jar is larger, but is this economy? One should quickly see that it is not. The 5 ounce jar at

35 cents means each ounce costs 7 cents; thus, 9 ounces at the *same* price would be 63 cents. Consequently, 65 cents for the 9 ounce jar would not be a saving. Let us consider two other buying problems.

*Examples*

1. If a $10\frac{3}{4}$ ounce package of a given product is selling for 42 cents and the $5\frac{5}{8}$ ounce package of the same product is selling for 22 cents, which is the better buy?
   *Solution.* Since
   $$42 \div 10\frac{3}{4} = 42 \div \frac{43}{4} = 42 \times \frac{4}{43} = \frac{168}{43} = 3\frac{39}{43},$$
   the large package costs $3\frac{39}{43}$ cents per ounce. Since
   $$22 \div 5\frac{5}{8} = 22 \div \frac{45}{8} = 22 \times \frac{8}{45} = \frac{176}{45} = 3\frac{41}{45},$$
   the small package costs $3\frac{41}{45}$ cents per ounce. Now,
   $$\frac{39}{43} < \frac{41}{45},$$
   since
   $$39 \times 45 < 43 \times 41,$$
   or
   $$1755 < 1763.$$
   Therefore, the large package is a better buy.

2. Suppose a room is 9 feet wide and 11 feet 3 inches long. If the floor is to be tiled with tiles which are 9 inches square and which cost 22 cents each, what is the cost of the tile needed to cover the floor?
   *Solution.* Since 9 feet is $9 \times 12 = 108$ inches and since 11 feet 3 inches is $(11 \times 12) + 3 = 135$ inches, the floor contains $108 \times 135 = 14{,}580$ square inches. Since each tile is a 9 inches square, each tile contains $9 \times 9 = 81$ square inches. Thus, $14{,}580 \div 81 = 180$ tiles are needed to cover the floor. Since each tile costs 22 cents, the total cost of the 180 tiles is $22 \times 180 = 3960$ cents, or $39.60.

11.6 *Miscellaneous Applications* 187

Recall that we may add or subtract the same number from both sides of an inequality and we may multiply or divide both sides of an inequality by a positive number without changing the sense of the inequality. With these facts at hand, we can solve some other problems of interest.

*Examples*

1. Suppose the total number of units that can be produced in a given day by Mr. Brown and Mr. Smith working at separate machines is 100. If it is known that Mr. Brown working 5 days and Mr. Smith working 3 days can produce no more than 424 units of the product, what is the most number of units that could be produced by Mr. Brown in one day? (We assume each man produces the same number each day.)

   *Solution.* If $a$ is the number of units Mr. Brown produces in one day, then, since he and Mr. Smith produce 100 units in one day, the number of units which Mr. Smith produces in one day is $100 - a$. In 5 days Mr. Brown will produce $5a$ units and in 3 days Mr. Smith will produce $3(100 - a)$. Since the total number is no more than 424 units, we have

   $$5a + 3(100 - a) \leq 424,$$

   $$5a + 300 - 3a \leq 424,$$

   $$5a - 3a \leq 424 - 300,$$

   $$2a \leq 124,$$

   $$a \leq 62.$$

   Thus, the maximum number that can be produced by Mr. Brown is 62.

2. Two trucks must make deliveries totaling 100 miles. Truck A gets 15 miles per gallon of fuel and truck B gets 10 miles per gallon. If the total fuel consumption is to be no more than 8 gallons, what is the minimum number of miles truck A must travel?

   *Solution.* If $a$ is the distance truck A travels, then $100 - a$ is the distance truck B travels. Since truck A can travel 15 miles on one gallon, for $a$ miles it uses $\frac{a}{15}$ gallons. Similarly, truck B uses $\frac{100-a}{10}$ gallons. Now, since the total fuel consumption must be no more than 8 gallons, we must have

$$\frac{a}{15} + \frac{100-a}{10} \leq 8.$$

Thus, multiplying both sides by 30 we get

$$2a + 300 - 3a \leq 240,$$
$$300 - 240 \leq 3a - 2a,$$
$$60 \leq a.$$

Hence, truck A must travel at least 60 miles so that total fuel consumption will be no more than 8 gallons.

## Exercises I

1. A $3\frac{3}{8}$ ounce jar of a product sells for 38 cents and a 7 ounce jar of the same product sells for 79 cents. Which is the better buy?
2. Three cans of a product weighing $5\frac{1}{4}$ ounces per can sells for 89 cents. If a large can of the same product weighs 14 ounces and sells for 79 cents a can, which is the better buy?
3. A room is 11 ft 6 inches wide and 18 ft long. If the floor is to be tiled with tiles which are 9 inches square and cost $19\frac{1}{2}$ cents each, what is the cost of the tile needed to cover the floor?
4. A gallon of ceiling paint is said to cover 450 sq ft. Suppose the ceilings in three rooms with dimensions 9 ft by 12 ft 3 in., 10 ft by 11 ft 6 in., and 13 ft 6 in. by 18 ft 2 in. are to be painted with this paint. How many quarts will be needed?
5. If gasoline costs 38.9 cents per gallon and if a 300 mile trip is to be made in a car that averages $15\frac{1}{2}$ miles per gallon, what will be the cost of the gasoline needed to make the trip?
6. Suppose car A averages 14.2 miles per gallon using gasoline costing 38.9 cents per gallon and suppose car B averages

17.6 miles per gallon using gasoline costing 34.9 cents per gallon. If both cars are driven 10,000 miles, what is the difference in the total costs of the amounts of gasoline needed by the two cars.
7. Assume a home has a rectangular front lawn with dimensions 30 ft by 80 ft and a rectangular back lawn with dimensions 55 ft by 95 ft. If 40 pounds of lawn food properly applied covers 10,000 sq ft and if it costs $0.30 per pound, what is the total cost of the lawn food necessary for proper application to the front and back yards?
8. A 54 pound bag of lawn food which is intended to cover 5,000 sq ft costs $13.95 for each bag. How much would a proper application of this lawn food cost if it were applied to the yards described in Exercise 7?
9. The total cost of a mower of type A and a mower of type B is $486. If an order with 5 mowers of type A and 2 mowers of type B costs less than $1,800, what is the maximum possible price for a mower of type A?
10. Two trucks must make deliveries totaling 300 miles. Truck A gets 15 miles per gallon of fuel and truck B gets 8 miles per gallon. If the total fuel consumption is to be no more than 30 gallons, what is the minimum number of miles that truck A must travel?

## 11.7 Arithmetic Sequences

Consider the set of odd positive integers: 1, 3, 5, 7, 9, 11, 13, . . . . What is the 50th term in this sequence of odd integers? The second term is obtained by adding two to the first term; the third term is obtained by adding two twos to the first term; the fourth term is obtained by adding three twos to the first term; and the 50th term is obtained by adding 49 twos to the first term. Thus, the 50th odd positive integer is $1 + (49)2 = 99$. In general, the $n$th term of the sequence is $1 + (n - 1)2$.

Suppose we wish to find the sum $S$ of the first 50 odd integers. This can be done in several ways. One way, though not very efficient, is to copy down in a column all the odd numbers from 1 to 99 and add; besides the possibility of a mistake in addition, just writing them all down would not be a very pleasant task. A second and

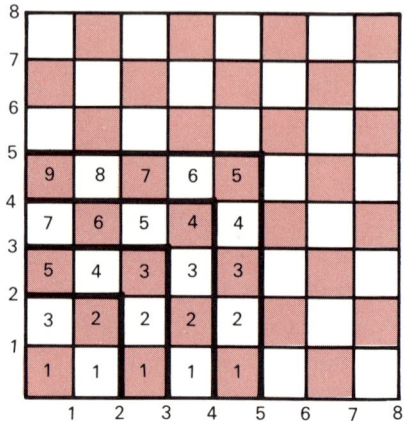

Figure 11.  $1 + 3 + 5 + 7 + 9 = 5^2$.

more efficient method would be to look at a tile floor. (See Figure 11.) If we start in the lower left-hand corner and add the 3 adjacent squares to the first square, then we get a new square that is 2 by 2. Thus,

$$1 + 3 = 2^2.$$

To get the next larger square which is 3 by 3, we add the 5 adjacent tiles to the preceding square. Thus,

$$1 + 3 + 5 = 3^2.$$

Likewise, $1 + 3 + 5 + 7 = 4^2$, $1 + 3 + 5 + 7 + 9 = 5^2$, and $1 + 3 + 5 + 7 + 9 + 11 = 6^2$. Continuing, we conclude that

$$1 + 3 + 5 + \cdots + 97 + 99 = 50^2.$$

Hence, the sum of the first 50 odd integers is 2500. In general, the sum of the first $n$ odd integers is $n^2$

A third way to find this sum uses the fact that the order in which numbers are added does not affect the sum. Hence, if

$$S = 1 + 3 + 5 + \cdots + 95 + 97 + 99, \qquad (1)$$

then

$$S = 99 + 97 + 95 + \cdots + 5 + 3 + 1. \qquad (2)$$

Using the fact that if $A = B$ and $C = D$ then $A + C = B + D$, we add the two equalities (1) and (2) and get

$$2S = 100 + 100 + 100 + \cdots + 100 + 100 + 100.$$

Since there are fifty terms each of which is 100,
$$2S = 50(100)$$
and
$$S = (25)(100) = 2500.$$

This third technique for finding the sum of the first 50 odd integers is quite significant since it lends itself to generalization. Suppose we wish to find the sum of the first 100 positive integers. If
$$T = 1 + 2 + 3 + \cdots + 97 + 98 + 99 + 100,$$
then
$$T = 100 + 99 + 98 + \cdots + 4 + 3 + 2 + 1.$$
Adding the equalities,
$$2T = \underbrace{101 + 101 + 101 + \cdots + 101 + 101 + 101 + 101}_{100 \text{ terms}};$$
thus,
$$2T = 100(101),$$
$$T = 50(101) = 5050.$$

Could this last technique be used to find the sum of the first 50 perfect squares? To answer this question, let us try it on the sum of just the first 5 perfect squares.

If
$$S = 1^2 + 2^2 + 3^2 + 4^2 + 5^2,$$
then
$$S = 5^2 + 4^2 + 3^2 + 2^2 + 1^2;$$
that is,
$$S = 1 + 4 + 9 + 16 + 25$$
and
$$S = 25 + 16 + 9 + 4 + 1.$$
Adding the last two equalities, we get
$$2S = 26 + 20 + 18 + 20 + 26.$$

Obviously, not all of the terms are the same and our new found technique "breaks down." But on what kind of sums does the technique work? In other words, what will ensure that when you write down the terms of some given sequence of numbers in reverse order and add the corresponding terms of the original sequence with the new sequence that the terms of the resulting sum are all the same? A careful study of the situation will show that this will be guaranteed if the difference between every pair of consecutive terms is the same. Such a sequence of terms is called an arithmetic sequence (progression).

For an arithmetic sequence with first term $a$, the second term is $a + d$ where $d$ is the common difference between the terms. The third term is $a + 2d$, the fourth term is $a + 3d$, the fifth is $a + 4d$, and the $n$th term is $a + (n - 1)d$. For example, 4, 7, 10, 13, 16, ... is an arithmetic sequence where the first term $a$ is 4 and the common difference $d$ is 3; furthermore, the 14th term is $4 + (13)3 = 4 + 39 = 43$.

A formula for the sum of an arithmetic sequence can be found by using the same technique we used in our examples. The sum $S$ of $n$ terms is

$$S = a + [a + d] + [a + 2d] + \cdots + [a + (n - 2)d] + [a + (n - 1)d];$$

thus, writing the terms of the sum in reverse order we get

$$S = [a + (n - 1)d] + [a + (n - 2)d] + [a + (n - 3)d] + \cdots + [a + d] + a.$$

Adding the last two equalities,

$$2S = \underbrace{[2a + (n - 1)d] + [2a + (n - 1)d] + \cdots + [2a + (n - 1)d]}_{n \text{ terms}};$$

hence,

$$2S = n[2a + (n - 1)d]$$

and

$$S = \frac{n}{2}[2a + (n - 1)d].$$

Consequently, $S = (\frac{n}{2})[2a + (n - 1)d]$ is the formula for the sum of $n$ terms of any arithmetic sequence with the first term $a$ and common difference $d$.

## 11.7 Arithmetic Sequences

### Examples

1. Suppose a man secures a job with a starting salary of $10,000 a year. If he gets a $600 raise each year, what are his total earnings for 20 years?

   *Solution.* This is an arithmetic sequence where $a = 10,000$, $d = 600$, and $n = 20$. Thus,

   $$S = \frac{20}{2}[20,000 + (19)(600)]$$
   $$= 10[20,000 + 11,400]$$
   $$= 314,000 \text{ dollars}$$

   is his total income for the twenty-year period.

2. If a person saves five cents on a given day, ten cents the next day, fifteen cents the third, twenty cents the fourth, etc. and maintains this savings program for thirty days, what is the total amount saved?

   *Solution.* The daily savings form an arithmetic sequence where $a = 5$, $d = 5$, and $n = 30$. Hence, the total savings $S$ is

   $$S = \frac{30}{2}[2(5) + (29)5]$$
   $$= 15[10 + 145]$$
   $$= 2325 \text{ cents.}$$

   Hence, the total amount saved is $23.25.

3. The *arithmetic mean (average)* of a set of numbers is the sum of the numbers in the set divided by the number of numbers in the set. Show that the sum of an arithmetic sequence is the product of the number of terms and the arithmetic mean of the first and last term.

   *Solution.* Since the first term is $a$ and the last term is $a + (n-1)d$, the arithmetic mean of the first and last term is

   $$\frac{a + a + (n-1)d}{2} = \frac{2a + (n-1)d}{2}.$$

   Since the number of terms is $n$,

$$n \times \frac{2a + (n-1)d}{2} = \frac{n}{2}[2a + (n-1)d],$$

which is the sum of the arithmetic sequence.

4. (a) If the grades made on the first five tests a student takes are 68, 97, 84, 78, and 93, what is his grade average? (b) If the student takes another test and if all six tests are weighted equally, is it possible for him to raise his average by 3 points if 100 is the highest grade that can be made on the sixth test?

*Solution.* (a) His average on the first five tests is

$$\frac{68 + 97 + 84 + 78 + 93}{5} = \frac{420}{5} = 84.$$

(b) The student needs to obtain a grade of $x$ such that $x \leq 100$ and

$$\frac{68 + 97 + 84 + 78 + 93 + x}{6} = 84 + 3;$$

that is,

$$420 + x = 87(6),$$

$$x = 522 - 420,$$

$$x = 102.$$

He cannot raise his grade to an average of 87.

We can easily prove with our formula that the sum of the first $n$ positive integers is $n(n + 1)/2$. The first $n$ positive integers is an arithmetic sequence where $a = 1$, $d = 1$ and the number of terms is $n$. If $S$ is the sum, then

$$S = \frac{n}{2}[2 + (n-1)(1)]$$

$$= \frac{n}{2}[2 + n - 1]$$

$$= \frac{n(n+1)}{2}.$$

Let us give a geometric argument for finding the sum of the first

## 11.7 Arithmetic Sequences

six positive integers. It should be clear that the procedure can be generalized for any positive integer $n$. On a piece of rectangular graph paper place dots as in Figure 12(a). If $S$ is the number of dots, then it should be clear by observing how many dots are on each line segment that $S = 1 + 2 + 3 + 4 + 5 + 6$. In Figure 12(b), we notice that the total number of solid dots is also $S$; thus the total number of points indicated is $2S$. Obviously, the configuration in Figure 12(b) is a rectangle consisting of 6 rows with 7 points in each row. Hence, the total number of points in the configuration is $(6)(7)$. Consequently, $2S = (6)(7)$ and $S = \frac{(6)(7)}{2}$.

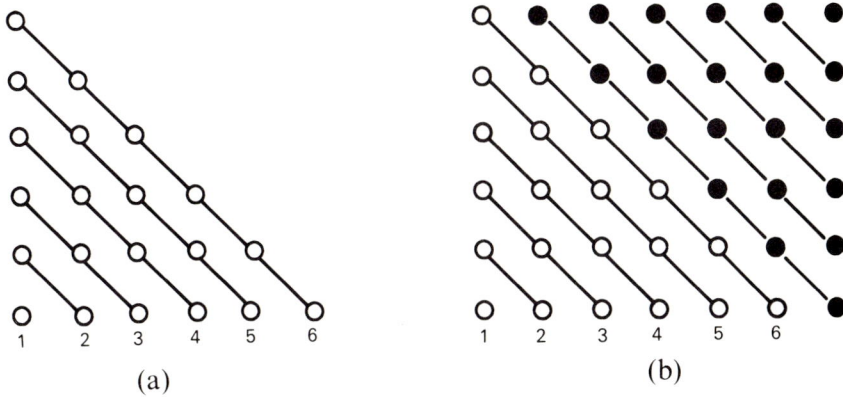

Figure 12. (a) $S = 1 + 2 + 3 + 4 + 5 + 6$. (b) $2S = (6)(7)$, $S = \frac{(6)(7)}{2} = 21$.

### Exercises I

1. (a) In an arithmetic sequence with the first term 4 and common difference 5, what is the 50th term?
   (b) What is the sum of the first 50 terms?
2. (a) If 5, 8, 11, ... , 65 is an arithmetic sequence, what is the common difference?
   (b) How many terms are there in the sequence if 65 is the last term?
   (c) What is the sum of the sequence.
   (d) Show for any arithmetic sequence with last term $l$ that the sum $S$ is given by $S = (\frac{n}{2})(a + l)$.
3. (a) In an arithmetic sequence with first term 3 and common difference $\frac{4}{3}$, what is the 50th term?

(b) What is the sum of the first 50 terms?
4. In an arithmetic sequence, if the first term is 5 and the 31st term is 98, what is the common difference?
5. For each of the following arithmetic sequences, answer the given questions.
   (a) 8, 11, 14, .... What is $d$? What is $S$ if $n = 30$? What is the 50th term?
   (b) 5, $y$, 12, .... What is $y$? What is $d$? What is $S$ if $n = 20$?
6. For each of these arithmetic sequences, answer the given questions:
   (a) 4, 7, 10, ..., 175. What is $d$? What is $n$? What is $S$?
   (b) 2, $\frac{10}{3}$, .... What is $d$? What is the third term? What is the 27th term?
7. For each of these arithmetic sequences, answer the given questions.
   (a) 8, $s$, $t$, $u$, 18, .... What is $s$? What is $t$? What is $u$? What is $S$ if $n = 10$?
   (b) $-3$, 1, 5, ..., 97. What is $d$? What is $n$? What is $S$?
8. (a) For the arithmetic sequence 4, $u$, $v$, $-2$, ..., what is $u$? What is $v$? What is $S$ if $n = 20$?
   (b) If 2 is the first term of an arithmetic sequence and 52 is the 21st, what is $d$? What is the sum of the first 21 terms?
9. A man secures a job with a starting salary of $8,300. If he receives a $500 raise each year, how much would his salary be during the 15th year on the job?
10. John's average for two tests is 86. What grade would he need to make on a third test so that his average for all three tests would be at least 90?

## 11.8 Geometric Sequences

What is a large number? The concept of bigness and smallness in numbers is quite relative and the answer to this question would depend on the position of the observer. Suppose the distance from the earth to the sun is measured as 193,000,000 miles and suppose this is incorrect by 100,000 miles. The percentage of error in this measurement is only about one-half the percentage of error in the

## 11.8 Geometric Sequences

length of a yard stick that is 36.04 inches long. From the viewpoint of percentage of error, 0.04 inches is a bigger error than 100,000 miles.

How big is the national debt of the U.S.A.? In relation to the Gross National Product, some would say that it is not so very large. However, suppose we look at the size of the debt from a different viewpoint. In November 1969, the national debt was somewhat more than \$368,000,000,000 (three hundred sixty-eight billion dollars). If a person were to count a dollar a second, sixty seconds each minute, sixty minutes each hour, twenty-four hours each day, and three hundred sixty-five days each year without stopping to eat, sleep, etc., how long would it take to count the national debt? Ten minutes? Ten hours? Ten days? Perhaps ten years? Take a guess.

Since there are 60 seconds in a minute and 60 minutes in an hour, there are $60 \times 60 = 3600$ seconds in an hour. Since there are 24 hours in a day, there are $24 \times 3600 = 86,400$ seconds in a day and $365 \times 86,400 = 31,536,000$ seconds in a (non-leap) year. Thus, in one year, a person could count \$31,536,000 at a dollar a second. Since the national debt divided by the number of dollars that can be counted per year is

$$368,000,000,000 \div 31,536,000 = 11,669 \text{ approx.},$$

it would take in excess of 11,669 years to count the national debt. It would be necessary to have started more than nine thousand years before the birth of Christ to have finished the counting by November 1969.

Another problem which has to do with the size of numbers is the following. If Mr. Smith were paid a salary of \$10,000 a day for 31 days and Mr. Jones earned 1 cent the first day, 2 cents the second day, 4 cents the third day, 8 cents the fourth day, 16 cents the fifth day, 32 cents the sixth day, and $2^{n-1}$ cents the $n$th day, which thirty-one day total salary is better? Since we bother to ask the question one might guess that Mr. Jones has the better salary. How much do you think Mr. Smith would have to make each day to have the same total salary as Mr. Jones? To determine how much Mr. Smith needs to make each day, we must find what Mr. Jones' total salary would be.

If $S$ is Mr. Jones' salary, then

$$S = 1 + 2 + 2^2 + 2^2 + \cdots + 2^{29} + 2^{30} \text{ cents.} \qquad (1)$$

Of course, one could find this sum by adding the thirty-one given numbers. However, the following procedure makes this unnecessary. Multiplying both sides of Equation (1) by 2, we obtain

$$2S = 2 + 2^2 + 2^3 + 2^4 + \cdots + 2^{30} + 2^{31}. \qquad (2)$$

Subtracting, $S = 1 + 2 + 2^2 + 2^3 + \cdots + 2^{29} + 2^{30}$ from Equation (2), we get

$$S = 2^{31} - 1 \text{ cents.}$$

Since $2^4 = 16, 2^8 = (2^4)^2 = (16)^2 = 256, 2^{16} = (2^8)^2 = 256 \times 256 = 65{,}536$ and $2^{31} = (2^{15})(2^{16}) = 32{,}768 \times 65{,}536 = 2{,}147{,}483{,}648$, it follows that

$$(2^{31} - 1) \text{ cents} = 2{,}147{,}483{,}647 \text{ cents} = \$21{,}474{,}836.47.$$

Since $21,474,836.47 divided by 31 is approximately $692,736.66, Mr. Smith would need to make $692,736.66 a day (more than two-thirds of a million dollars a day) in order to have a thirty-one day salary equal to Mr. Jones!

The salary schedule for Mr. Jones forms what is called a geometric sequence. In general, if $a$ is the first term of a sequence and if each successive term is obtained by multiplying the preceding term by a fixed number $r$, then the sequence is called *geometric*. The number $r$ is called the common ratio. In a geometric sequence, $a$ is the first term, $ar$ is the second term, $ar^2$ is the third term, $ar^3$ is the fourth term, and the $n$th term is $ar^{n-1}$. In the geometric sequence

$$1, 2, 2^2, 2^3, 2^4, \ldots, 2^{29}, 2^{30},$$

$a = 1, r = 2$, and $n = 31$. The procedure used above to find the sum of this geometric sequence can be used on any geometric sequence with a finite number of terms.

Consider the geometric sequence $a, ar, ar^2, ar^3, \ldots, ar^{n-2}, ar^{n-1}$ with $n$ terms and let

$$S = a + ar + ar^2 + ar^3 + \cdots + ar^{n-2} + ar^{n-1}. \qquad (1)$$

Multiplying both sides of Equation (1) by the common ratio $r$, we obtain

$$rS = ar + ar^2 + ar^3 + ar^4 + \cdots + ar^{n-1} + ar^n. \qquad (2)$$

Subtracting Equation (1) from Equation (2), we get

$$rS - S = ar^n - a,$$
$$S(r - 1) = ar^n - a,$$

## 11.8 Geometric Sequences

and

$$S = \frac{ar^n - a}{r - 1} \quad \text{provided } r \neq 1.$$

If $r = 1$, each of the $n$ terms of the sequence is $a$ and the sum is $na$.

### Examples

1. For the geometric sequence $\frac{2}{3}, \frac{1}{3}, \frac{1}{6}, \ldots$, find the common ratio and the sixth term.

   *Solution.* Since we are given that the sequence is geometric, we can find the common ratio by dividing any given term by the preceding term. Thus,

   $$r = \frac{1}{3} \div \frac{2}{3} = \frac{1}{3} \times \frac{3}{2} = \frac{1}{2}.$$

   The sixth term is

   $$\frac{2}{3}\left(\frac{1}{2}\right)^5 = \frac{2}{3} \times \frac{1}{32} = \frac{1}{48}.$$

2. For the geometric sequence 3, 3.6, 4.32, ..., find the common ratio and the sixth term.

   *Solution.* We may find the ratio in decimal or in fractional form. Since some rationals cannot be expressed as finite decimals, we usually choose to find a fractional expression for the ratio. Thus,

   $$r = 3.6 \div 3 = \frac{36}{10} \div 3 = \frac{18}{5} \times \frac{1}{3} = \frac{6}{5}.$$

   The sixth term is

   $$3\left(\frac{6}{5}\right)^5 = \frac{23{,}328}{3{,}125}.$$

A geometric sequence is obtained when money is deposited at compound interest. Suppose one deposits $P$ dollars at an interest rate of 6 percent per year. At the end of one year the interest is $0.06P$. The total amount on deposit will be the original principal $P$ plus the interest; that is, the total amount on deposit is

$$P + 0.06P = (1.06)P.$$

If the money is left on deposit, then at the end of the second year the amount of money on deposit is the total on deposit at the end of the first year plus the second year's interest; that is,

$$(1.06)P + (0.06)[(1.06)P] = (1.06)P[1 + 0.06]$$
$$= (1.06)P(1.06)$$
$$= (1.06)^2 P.$$

At the end of three years, the amount is

$$(1.06)^3 P,$$

and at the end of $n$ years it is

$$(1.06)^n P.$$

Notice that $P$, $(1.06)P$, $(1.06)^2 P$, ... is a geometric sequence with first term $P$ and common ratio $(1.06)$.

You might like to know how long it takes to double your money if it is deposited at 6 percent interest and compounded annually. This involves determining the number $n$ such that $(1.06)^n P = 2P$, or

$$(1.06)^n = 2.$$

The answer can be found easily in interest tables. (See back of book for interest tables.) In an interest table, you will find that

$$(1.06)^{12} = 2.01219647;$$

thus, in twelve years you would double your money with this type of investment.

In general, the amount $A$ after $n$ years of compound interest on a principal at an $i$ interest rate is given by

$$A = P(1 + i)^n.$$

It can also be proved that if interest is compounded $q$ times per year that the amount is given by

$$A = P\left(1 + \frac{i}{q}\right)^{qn}.$$

### Example

Suppose $5,000 is deposited at 5 percent interest compounded quarterly. If no money is added or withdrawn from the account,

## 11.8 Geometric Sequences

what would be the total amount on deposit at the end of 5 years?

Solution. $P = 5{,}000$, $i = 0.05$, $q = 4$, and $n = 5$. Thus,

$$A = 5{,}000\left(1 + \frac{0.05}{4}\right)^{20}$$

$$= 5{,}000(1 + 0.0125)^{20}.$$

From the interest tables in the back of the book, we see that

$$(1 + 0.0125)^{20} = 1.28203723.$$

Hence,

$$5{,}000(1.28203723) = 6{,}410.18615000,$$

and the amount at the end of 5 years is $6,410.18.

## Exercises I

In Exercises 1 through 6, find the common ratio and the seventh term of the given geometric sequences.

1. $3, -3, 3, \ldots$
2. $60, 30, 15, \ldots$
3. $2, 6, 18, \ldots$
4. $4, -2, 1, \ldots$
5. $81, -27, 9, \ldots$
6. $8, 4.8, 2.88, \ldots$
7. (a) If $1000 is invested with a 5 percent interest rate which is compounded quarterly, what is the value of the investment at the end of the year?
   (b) What simple interest would be equivalent to the 5 percent interest compounded quarterly?
8. How many years would it take to double your investment if money were deposited at 6 percent compounded quarterly?
9. How many years would it take to double your investment if money were deposited at 5 percent compounded quarterly?
10. How many years would it take to double your investment if money were deposited at 3 percent compounded quarterly?
11. How long must a given sum of money be invested at 5 percent interest compounded annually in order to receive the equal return of a ten-year investment at 3 percent interest rate compounded quarterly?

**12**

$a \equiv b, \bmod m, b \equiv a, \bmod m$

| × | 0 | 1 |
|---|---|---|
| 0 | 0 | 0 |
| 1 | 0 | 1 |

0.123123123

$1/7 + 1/14 + 1/28$

# BASIC THEOREMS

## 12.1 Divisibility Tests

In Chapter 5 we discussed several divisibility tests; let us now restate them and prove the validity of each test for a six-digit number represented by $abc,def$. The basic features of a general proof for an $n$-digit number would be the same; however, the proof would be rather tedious to write out.

**Test 1.** A positive integer is divisible by 2 if and only if the digit in the units position is 0, 2, 4, 6, or 8.
  *Example.* 238 is divisible by 2, and 769 is not divisible by 2.

*Proof of Test 1.* (We need to prove two theorems, an implication and its converse.) Part 1. Let $abc,def$ be a six-digit number where $f$ is 0, 2, 4, 6, or 8. As a consequence of the positional notation,
$$abc,def = 10^5 a + 10^4 b + 10^3 c + 10^2 d + 10e + f.$$
Hence,
$$\begin{aligned} abc,def &= 10(10)^4 a + 10(10)^3 b + 10(10)^2 c + 10(10)d + 10e + f \\ &= 2[5(10)^4 a + 5(10)^3 b + 5(10)^2 c + 5(10)d + 5e] + f. \end{aligned}$$
Thus, $abc,def$ is the sum of the two numbers
$$2[5(10)^4 a + 5(10)^3 b + 5(10)^2 c + 5(10)d + 5e] \quad \text{and} \quad f.$$
The first number obviously has 2 as a factor. Since $f$ is 0, 2, 4, 6, or 8, $f$ has 2 as a factor. (We recall that if two numbers have a factor in common then so does the sum of the two numbers.) Hence, 2 is a factor of the given number $abc,def$.

*Part 2.* Assume 2 is a factor of a six-digit number $abc,def$. From Part 1, we have

$$f = abc,def - 2[5(10)^4 a + 5(10)^3 b + 5(10)^2 c + 5(10)d + 5e].$$

Since 2 is a factor of $abc,def$ and since 2 is a factor of $2[5(10)^4 a + 5(10)^3 b + 5(10)^2 c + 5(10)d + 5e]$, we know that 2 is a factor of the difference of the two numbers; that is, 2 is a factor of $f$. Since $f$ must be 0, 1, 2, 3, 4, 5, 6, 7, 8, or 9 and since of these 2 is a factor of 0, 2, 4, 6, or 8 only, it follows that $f$ must be one of 0, 2, 4, 6, or 8.

**Test 2.** A positive integer is divisible by 3 if and only if the sum of the digits is divisible by 3.

*Example.* 472,641 is divisible by 3 since $4 + 7 + 2 + 6 + 4 + 1 = 24$ is divisible by 3. The number 64,534 is not divisible by 3 since $6 + 4 + 5 + 3 + 4 = 22$ is not divisible by 3.

*Proof of Test 2. Part 1.* Let $abc,def$ be a six-digit number and assume 3 is a factor of $a + b + c + d + e + f$. (We need to prove that 3 is a factor of $abc,def$.) Since

$$abc,def = 10^5 a + 10^4 b + 10^3 c + 10^2 d + 10e + f,$$

it follows that

$$\begin{aligned} abc,def &= 100{,}000a + 10{,}000b + 1{,}000c + 100d + 10e + f \\ &= (99{,}999 + 1)a + (9{,}999 + 1)b + (999 + 1)c + \\ &\qquad (99 + 1)d + (9 + 1)e + f \\ &= (99{,}999a + 9{,}999b + 999c + 99d + 9e) + \\ &\qquad (a + b + c + d + e + f) \\ &= 3(33{,}333a + 3{,}333b + 333c + 33d + 3e) + \\ &\qquad (a + b + c + d + e + f). \end{aligned}$$

Since 3 is a factor of both $(a + b + c + d + e + f)$ and $3(33{,}333a + 3{,}333b + 333c + 33d + 3e)$, it is a factor of the sum of the two numbers; that is, 3 is a factor of $abc,def$.

*Part 2.* Left as an exercise.

**Test 3.** A positive integer greater than 10 is divisible by 4 if and only if the number represented by the digits in the tens' and units' places is divisible by 4.

*Example.* 378,569,532 is divisible by 4 since 32 is divisible by 4. Also, 2,576,426 is not divisible by 4 since 26 is not divisible by 4.

## 12.1 Divisibility Tests

*Proof of Test 3. Part 1.* Let $abc,def$ be a six-digit number and assume $ef$ is divisible by 4. Since

$$abc,def = 10^5 a + 10^4 b + 10^3 c + 10^2 d + 10e + f,$$

it follows that

$$abc,def = (100{,}000a + 10{,}000b + 1{,}000c + 100d) + (10e + f)$$
$$= 4(25{,}000a + 2{,}500b + 250c + 25d) + (10e + f).$$

Since 4 is a factor of $ef$, 4 is a factor of $(10e + f)$. Consequently, 4 is a factor of the sum

$$4(25{,}000a + 2{,}500b + 250c + 25d) + (10e + f);$$

that is, 4 is a factor of $abc,def$.
*Part 2.* Left as an exercise.

**Test 4.** A positive integer is divisible by 5 if and only if the digit in the units' position is 0 or 5.
*Proof of Test 4.* Left as an exercise.

**Test 5.** A positive integer is divisible by 6 if and only if the integer is even and the sum of the digits is divisible by 3.
*Proof of Test 5.* Left as an exercise.

**Test 6.** A positive integer greater than 100 is divisible by 8 if and only if the number represented by the digits in the hundreds', tens', and units' positions is divisible by 8.
*Proof of Test 6.* Left as an exercise.

**Test 7.** A positive integer is divisible by 9 if and only if the sum of the digits is divisible by 9.
*Proof of Test 7.* Left as an exercise.

**Test 8.** A positive integer is divisible by 10 if and only if the digit in the units' position is 0.
*Proof of Test 8.* Left as an exercise.

**Test 9.** A positive integer is divisible by 11 if and only if the difference of the sum of the digits in the odd-numbered positions (from the right) and the sum of the digits in the even-numbered positions is divisible by 11.
   *Example.* 92,818 is divisible by 11 since $(8 + 8 + 9) - (1 + 2) =$

22 is divisible by 11. Also, 867,534,281 is not divisible by 11 since $(1 + 2 + 3 + 7 + 8) - (8 + 4 + 5 + 6) = -2$ does not have 11 as a factor.

*Proof of Test 9. Part 1.* Let $abc,def$ be a six-digit number and assume that $(b + d + f) - (a + c + e)$ has 11 as a factor. Since

$$abc,def = 10^5 + 10^4 b + 10^3 c + 10^2 d + 10e + f,$$

it follows that

$$\begin{aligned}
abc,def &= 100{,}000a + 10{,}000b + 1{,}000c + 100d + 10e + f \\
&= (100{,}001 - 1)a + (9{,}999 + 1)b + (1{,}001 - 1)c + \\
&\qquad (99 + 1)d + (11 - 1)e + f \\
&= 100{,}001a + 9{,}999b + 1{,}001c + 99d + 11e - \\
&\qquad a + b - c + d - e + f \\
&= 11(9{,}091a + 909b + 91c + 9d + e) + \\
&\qquad [(b + d + f) - (a + c + e)].
\end{aligned}$$

Since the difference $(b + d + f) - (a + c + e)$ has 11 as a factor and since $11(9{,}091a + 909b + 91c + 9d + e)$ has 11 as a factor, the sum $abc,def$ has 11 as a factor.

*Part 2.* Left as an exercise.

The proof of Test 9 is the only one that does not generalize easily for an $n$-digit number. However, in Section 12.5, we shall give a more simple proof which will generalize quite easily. As with many proofs in mathematics, the latter proof will be "simpler" only because we shall have more sophisticated mathematical "tools" at our disposal.

### Exercises I

1. Which of the numbers 2, 3, 4, 5, 6, 8, 9, 10, 11, 15, 18, and 40 are factors of 22,440?
2. Which of the numbers 2, 3, 4, 5, 6, 8, 9, 10, 11, 15, 18, and 40 are factors of 51,524?
3. Which of the numbers 2, 3, 4, 5, 6, 8, 9, 10, 11, 15, 18, and 40 are factors of 26,640?
4. Which of the numbers 2, 3, 4, 5, 6, 8, 9, 10, 11, 15, 18, and 40 are factors of 172,159?

## 12.2 Infinitude of Primes

5. Which of the numbers 2, 3, 4, 5, 6, 8, 9, 10, 11, 15, 18, and 40 are factors of 1,022,380?
6. Which of the numbers 2, 3, 4, 5, 6, 8, 9, 10, 11, 15, 18, and 40 are factors of 1,141,140?
7. Which of the numbers 2, 3, 4, 5, 6, 8, 9, 10, 11, 15, 18, and 40 are factors of 805,950,981?
8. Which of the numbers 2, 3, 4, 5, 6, 8, 9, 10, 11, 15, 18, and 40 are factors of 55,074,492?
9. Which of the numbers 2, 3, 4, 5, 6, 8, 9, 10, 11, 15, 18, and 40 are factors of 2,010,360?
10. Which of the numbers 2, 3, 4, 5, 6, 8, 9, 10, 11, 15, 18, and 40 are factors of 2,968,985,017?

### Exercises II

1. Complete the proof of Test 2.
2. Complete the proof of Test 3.
3. Prove Test 4.
4. Prove Test 5.
5. Prove Test 6.
6. Prove Test 7.
7. Prove Test 8.
8. Complete the proof of Test 9.

## 12.2 Infinitude of Primes

Let us prove that there is not a last prime. Although the proof is neither long nor complicated, it does require a familiarity with some subtleties of logic. The proof is an indirect proof (*reductio ad absurdum*). We demonstrate that if we assume there is a last prime then it leads to a contradiction. Therefore, the assumption is untenable and we conclude there is no last prime.

**Theorem.** The set of prime numbers is not a finite set.

*Proof.* Assume there is a last (largest) prime and let it be $t$. Now, consider the number $x$, which is the sum of 1 and the product of *all* the primes from 2 to $t$; that is, let

$$x = 1 + (2 \times 3 \times 5 \times 7 \times \cdots \times t).$$

Since $x$ is greater than the last prime $t$, we conclude that $x$ is a composite number and, hence, must have some prime $p$ as a factor.*

Now, $p$ must be one of the primes in the product $(2 \times 3 \times 5 \times 7 \times \ldots \times t)$ since there are no other primes. Next, using the fact that if two numbers have a common factor then their difference also has this number as a factor, we conclude that

$$x - (2 \times 3 \times 5 \times 7 \times \cdots \times t)$$

has $p$ as a factor. But, this difference is equal to 1, and $p$ is *not* a factor of 1. Since the assumption that there are a finite number of primes leads to a contradiction, it follows that the set is an infinite set.

If $p$ and $q$ are primes and if they differ by two, then they are called *twin primes*. For example, 3 and 5, 5 and 7, 11 and 13, 17 and 19, and 29 and 31 are twin primes. (See Table 1.) Although mathematicians have solved many deep and difficult problems concerning the set of primes, it remains an unsolved problem whether or not the set of twin primes is a finite set.

## Exercises I

1. Express each even natural number between 4 and 98 as the sum of two primes.
2. Can any even natural number be expressed as the sum of two primes in more than one way?
3. A famous unsolved problem, known as Goldbach's conjecture, states that every even positive integer greater than 4 can be expressed as the sum of two odd primes. Suppose Goldbach's conjecture were correct. Could you prove that every even number greater than 8 could be expressed as the sum of three primes?
4. Find at least 5 natural number replacements for $N$ such that $N^2 + 1$ is a prime number.
5. It is unknown whether or not the set of primes of the form

---

*Here we use the fact that any composite number can be expressed as the product of primes; see Section 12.3 for a proof of this statement.

## 12.2 Infinitude of Primes

Table 1

*Primes and Twin Primes Less Than 1,000*

| Number of Primes | | | | | | | | | |
|---|---|---|---|---|---|---|---|---|---|
| 25 | **2** **31** **73** | **3** 37 79 | **5** **41** 83 | **7** **43** 89 | **11** 47 97 | **13** 53 | **17** **59** | **19** **61** | **23** 67 | **29** **71** |
| 21 | **101** **151** **199** | **103** 157 | **107** 163 | **109** 167 | 113 173 | 127 **179** | 131 **181** | **137** **191** | **139** **193** | **149** **197** |
| 16 | **211** **269** | 223 **271** | **227** 277 | **229** **281** | 233 **283** | **239** 293 | **241** | 251 | 257 | 263 |
| 16 | **307** 367 | **311** 373 | **313** **379** | 317 383 | 331 389 | 337 397 | **347** | **349** | 353 | 359 |
| 17 | **401** **461** | 409 **463** | **419** 467 | **421** 479 | **431** 487 | **433** 491 | 439 499 | 443 | 449 | 457 |
| 14 | 503 577 | 509 587 | **521** 593 | **523** **599** | 541 | 547 | 557 | 563 | **569** | **571** |
| 16 | **601** **659** | 607 **661** | 613 673 | **617** 677 | **619** 683 | 731 691 | **641** | **643** | 647 | 653 |
| 14 | **701** **769** | 709 773 | 719 **787** | 727 797 | 733 | 739 | 743 | 751 | 757 | 761 |
| 15 | **809** 863 | **811** 877 | **821** **881** | **823** **883** | **827** 887 | 829 | 839 | 853 | **857** | **859** |
| 14 | **907** **977** | 911 983 | 919 **991** | 929 **997** | 937 | 941 | 947 | 953 | 967 | 971 |

$N^2 + 1$ is a finite set. How many primes are there of the form $N^2 - 1$?

6. List fifteen consecutive positive integers none of which is a prime.
7. Is $n^2 - n + 41$ a prime number for $n = 1, 2, 3, 4, 5, 6, 7, 8, 9,$ and $10$?
8. Is $n^2 - n + 41$ a prime number for every positive integer $n$?
9. Can each of the first thirty odd positive integers be expressed as the sum of a power of 2 and a prime number?
10. Can every odd positive integer be expressed as the sum of a power of 2 and a prime number? (Hint: Try the odd numbers between 2 and 50 and between 144 and 150.)

## 12.3 Fundamental Theorem of Arithmetic

Recall that an integer $a$ is a factor of an integer $b$ if and only if there exists an integer $x$ such that $ax = b$. If $a$ is a factor of $b$, we say that *a divides b* and denote this fact by $a|b$. We have stated and use the fact that if an integer $a$ is a factor of two integers $b$ and $c$ then $a$ is a factor of their sum. Let us prove this.

**Theorem.** If $a|b$ and $a|c$, then $a|(b + c)$.

*Proof.* If $a|b$ and $a|c$, then there exist integers $x$ and $y$ such that $ax = b$ and $ay = c$. (Although $x$ and $y$ might be the same number, this need not be the case; hence, it is essential to use different letters.)

From $ax = b$ and $ay = c$, we conclude that

$$ax + ay = b + c.$$

Thus,

$$a(x + y) = b + c.$$

Since the sum of two integers is an integer, $(x + y)$ is an integer. By definition of factor, it follows that $a|(b + c)$.

We now state some other familiar theorems. The proofs will be left as exercises.

1. If $a|b$ and $a|c$, then $a|(b - c)$.
2. If $a|b$ and $b|c$, then $a|c$.
3. If $a|b$ or $a|c$, then $a|bc$.
4. If $a|b$, then $a^2|b^2$.

In Section 12.2 we proved that the set of prime numbers was an infinite set. In the proof, we tacitly assumed that every composite number could be expressed as the product of prime numbers. Now, we shall prove this important theorem.

**Theorem.** Every positive integer $n$ greater than 1 is either a prime number or it is expressible as the product of prime numbers.

*Proof.* Let $S$ be the set of positive integers which are primes or which can be expressed as the product of primes. Obviously, all primes greater than or equal to 2 are in $S$. If every composite

## 12.3 Fundamental Theorem of Arithmetic

number is in $S$, then the theorem is true. (We shall prove this is the case by showing the assumption that there is a composite number not in $S$ leads to a contradiction.) Let $T$ be the set of composite numbers not in $S$. If $T$ is not empty, then there exists a least positive integer $k \in T$ by the well-ordering property. Since $k$ is a composite number, there exist at least two factors $a$ and $b$ of $k$ such that neither is equal to 1 and such that

$$k = ab.$$

Thus, $a < k$ and $b < k$. Since $k$ is the least element in $T$ and since $a \neq 1$ and $b \neq 1$, it follows that $a \in S$ and $b \in S$. Consequently, either $a$ is a prime or it is expressible as the product of primes; similarly, the same is true for $b$. Therefore, since $k$ is the product of $a$ and $b$, it can be expressed as the product of primes and therefore is in $S$. But this is a contradiction, showing that $T$ is empty, and the theorem is proved.

The preceding theorem proves that a composite number such as 1,680 can be expressed as the product of primes. Notice that

$$1{,}680 = 2 \times 2 \times 2 \times 2 \times 3 \times 5 \times 7.$$

If several persons expressed 1,680 as the product of primes, it is not obvious that each would have the "same answer." In other words, it is not obvious that the prime factorization of a composite number is unique, except for the order in which the primes are written. The fact that a composite number is expressible as the product of primes in *one and only one way* is called the Fundamental Theorem of Arithmetic. Let us prove this important theorem.

**Theorem (Fundamental Theorem of Arithmetic).** Any composite number is expressible as the product of primes in one and only one way.

*Proof. Part 1.* The preceding theorem proved that a composite number is expressible as the product of primes in at least one way.
*Part 2.* Let $T$ be the set of composite numbers which can be expressed as the product of primes in *more than* one way. (We would like to prove that $T = \emptyset$; this will be done by contra-

diction.) Assume $T \neq \emptyset$. Then there exists a least composite $k$ in $T$ such that

$$k = xyz \quad \text{and} \quad k = abc \ldots$$

are two different prime factorizations of $k$. Let us first demonstrate that since the two sets of prime factors are different, no *two* elements in each set of factors can be the same. If, for example, $x = a$, then

$$yz \ldots = bc \ldots$$

are two different factorizations into primes of a composite less than $k$; this is contrary to the assumption that $k$ was the least composite in $T$.

Now, let us assume that $x$ is the least prime in the product $xyz \ldots$; hence, $x^2 \leq xy \leq k$. Similarly, if $a$ is the least prime in the product $abc \ldots$ then $a^2 \leq k$. Since $a \neq x$, $ax < k$ and $k - ax$ is a positive integer less than $k$ and it is expressible as the product of primes in only one way. Thus, since $a$ and $x$ are factors of $k$ and since $a \neq x$, it follows that

$$k - ax = axuvw \ldots,$$

where $u$, $v$, and $w$ are primes. Thus,

$$k = ax(1 + uvw \ldots),$$

and $ax$ is a factor of $k$; this implies that

$$abc \ldots = ax(1 + uvw \ldots).$$

Hence,

$$bc \ldots = x(1 + uvw \ldots),$$

and $x$ is a factor of $bc \ldots$ Now, $bc \ldots$ is the unique prime factorization of a composite number less than $k$ for which no factor is equal to $x$; this is impossible. The assumption that $T$ is not empty leads to a contradiction; thus, $T = \emptyset$ and each composite number can be expressed as the product of primes in only one way.

From a logical point of view, Parts 1 and 2 of the preceding theorem have an interesting difference. Although the proof of Part 1 is based on the definition of prime number, the well-ordering property, and the multiplicative properties for integers, Part 2 cannot be proved solely with these properties.

## 12.3 Fundamental Theorem of Arithmetic

Consider the sequence 1, 4, 7, 10, 13, 16, 19, 22, 25, 28, .... The general term of this arithmetic sequence is given by $3n - 2$ where $n$ is a positive integer. Since

$$(3n - 2)^2 = 9n^2 - 12n + 4$$
$$= 9n^2 - 12n + 6 - 2$$
$$= 3(3n^2 - 4n + 2) - 2,$$

the square of any number in this sequence is of the form $3N - 2$ where $N$ is an integer and, therefore, is a number in the original sequence. In fact, let $3n - 2$ and $3m - 2$ be any two numbers in the sequence; then

$$(3n - 2)(3m - 2) = 9mn - 6m - 6n + 4$$
$$= 9mn - 6m - 6n + 6 - 2$$
$$= 3(3mn - 2m - 2n + 2) - 2.$$

Hence, the product of any two numbers in the sequence is in the sequence; that is, the set is closed with respect to multiplication. If a *pseudo-prime* is defined as any number in the sequence having only two distinct factors, then, for example, 4, 7, 10, 13, 19, 22, 25, 31, etc., are pseudo-primes. Let any other number, except 1, in the sequence be called $T$-composite. It is true that $T$-composites such as 16, 28, 40, etc. in the sequence are expressible as the product of pseudo-primes. However, the decomposition (factorization) into pseudo-primes is not unique as seen by the pseudo-prime decompositions of 100.

$$100 = 4 \times 25 = 10 \times 10.$$

### Exercises I

Express each number in Exercises 1 through 10 as a product of primes.

1. 888.
2. 432.
3. 924.
4. 693.
5. 715.
6. 1,732.
7. 2,004.
8. 3,024.
9. 5,274.
10. 20,202.

### Exercises II

1. Prove that if $a|b$ and $a|c$, then $a|(b - c)$.
2. Prove that if $a|b$ and $b|c$, then $a|c$.

3. (a) Prove that is $a|b$ or $a|c$, then $a|bc$.
   (b) Show by example that the converse of the theorem stated in part (a) is not true.
4. Prove that if $a|b$, then $a^2|b^2$.
5. Show that the sequence $3n - 1$ where $n$ is any positive integer contains no perfect squares. (Hint: The set of positive integers can be split into the following three disjoint sets: $\{3n\}$, $\{3n - 1\}$, and $\{3n - 2\}$.)
6. Prove that the arithmetic sequence 3, 7, 11, 15, 19, 23, . . . has no perfect squares. (Hint: See Exercise 5.)
7. Prove that every T-composite in the sequence 4, 7, 10, 13, . . . is expressible as the product of pseudo-primes.
8. Using the proof of Part 2 of the Fundamental Theorem of Arithmetic as a model, where would the proof "break down" in trying to prove that every T-composite in the sequence 4, 7, 10, 13, . . . could be expressed as the product of primes in only one way?

## 12.4 Euclidean Algorithm

As we stated in Chapter 5, if $a$ and $b$ are natural numbers and if $a \geq b$ then there exist whole numbers $q$ and $r$ where $0 < q$ and $0 \leq r < b$ such that

$$a = qb + r.$$

For example, in the long division of 25 by 4 the incomplete quotient is 6 and the remainder is 1. It is intuitively obvious that, in general, $q$ and $r$ do exist and are unique. However, we now prove this statement.

**Theorem.** If $a$ and $b$ are natural numbers and if $a \geq b$, then (1) there exist whole numbers $q$ and $r$ where $0 < q$ and $0 \leq r < b$ such that $a = qb + r$, and (2) the whole numbers $q$ and $r$ are unique.

*Proof. Part 1.* Let $S$ be the set of all multiples of $b$ which are greater than $a$. Since

$$(a + 1)b = ab + b > ab \geq a,$$

the multiple $(a + 1)b$ of $b$ is greater than $a$ and the set $S$ is

## 12.4 Euclidean Algorithm

not empty. As a consequence of the well-ordering property, we know that $S$ contains a least element; it is a multiple of $b$. Let $tb$ be the least number in $S$. Since $tb$ is the least number in $S$, $tb > a$ and $(t - 1)b \leq a$. Also, since $a \geq b$, we know that $t \neq 1$ for if $t = 1$ then $tb = b$ and $b > a$, a contradiction. Thus, $t > 1$ and $t - 1 > 0$; furthermore, $0 \leq a - (t - 1)b \leq b$. (In our example, $a = 25$, $b = 4$, $S = \{28, 32, 36, 40, \ldots\}$, and $tb = 28$; furthermore, $t = 7$, $t - 1 = 6$, and $0 \leq 25 - 6(4) < 6$.)

Since $a = (t - 1)b + a - (t - 1)b$, by letting $q = t - 1$ and $r = a - (t - 1)b$ we have integers $q$ and $r$ where $q > 0$, $0 \leq r < b$, and $a = qb + r$.

*Part 2.* (Since we have proved that whole numbers $q$ and $r$ exist, let us prove that they are unique.) Assume whole numbers $q_1$, $q_2$, $r_1$, and $r_2$ exist such that

$$a = q_1 b + r_1 \quad \text{where } q_1 > 0 \text{ and } 0 \leq r_1 < b$$

and

$$a = q_2 b + r_2 \quad \text{where } q_2 > 0 \text{ and } 0 \leq r_2 < b.$$

Thus,

$$a - r_1 = b q_1 \tag{A}$$

and

$$a - r_2 = b q_2. \tag{B}$$

If $r_1 \neq r_2$, we may assume $r_1 < r_2$ without loss of generality; hence, $0 < r_2 - r_1$. It is also true that $r_2 - r_1 < b$. Subtracting equation B from A, we get

$$r_2 - r_1 = b(q_1 - q_2).$$

Since $r_2 - r_1$ is positive, $b(q_1 - q_2)$ is a positive multiple of $b$; that is, $b(q_1 - q_2) \geq b$. However, since $r_2 - r_1 < b$, this implies that $b(q_1 - q_2) < b$, a contradiction. Hence, $r_1 = r_2$.

If $r_1 = r_2$, then $b(q_1 - q_2) = 0$. Thus, since $b \neq 0$ we have that $q_1 - q_2 = 0$ and $q_1 = q_2$.

Now, we exhibit again the Euclidean algorithm for finding the G.C.F. of two numbers. To find the G.C.F. of 426 and 768, we proceed as follows.

```
              1
       426 ⌡ 768
             426          1
             342 ⌡ 426
                   342           4
                    84 ⌡ 342
                         336      14
            G.C.F. =     6 ⌡ 84
                             84
                              0
```

Let us prove that this procedure always determines the G.C.F. of two numbers.

Let $a$ and $b$ be positive integers where $a > b$. Since the remainder is always less than the divisor in each step of the Euclidean algorithm, we know that the number of required divisions is finite. The steps (divisions) in the Euclidean algorithm are exhibited as follows.

Step 1.  $\qquad a = bq_1 + r_1 \qquad$ where $\quad 0 < r_1 < b$.

Step 2.  $\qquad b = r_1 q_2 + r_2 \qquad$ where $\quad 0 < r_2 < r_1$.

Step 3.  $\qquad r_1 = r_2 q_3 + r_3 \qquad$ where $\quad 0 < r_3 < r_2$.

$\cdots$ $\qquad\qquad\qquad \cdots$

Step $(n-1)$.  $\qquad r_{n-3} = r_{n-2} q_{n-1} + r_{n-1} \qquad$ where $0 < r_{n-1} < r_{n-2}$.

Step $n$.  $\qquad r_{n-2} = r_{n-1} q_n \qquad$ where $r_n = 0$.

We have assumed that in the $n$th Step the remainder is zero. Let us prove that the greatest common factor of $a$ and $b$ is indeed $r_{n-1}$.

*Proof.* Consider Step 1 where $a = bq_1 + r_1$. If $G$ is any factor of $a$ and $b$, then $G$ is a factor of $bq_1$ and also the difference $a - bq_1$; that is, $G$ is also a factor of $r_1$. Conversely, since any factor of $b$ and $r_1$ is a factor of $bq_1 + r_1$, it follows that any factor of $b$ and $r_1$ is a factor of $a$. Consequently, $G$ is the *greatest* common factor of $a$ and $b$ if and only if it is the greatest common factor of $b$ and $r_1$.

Similarly, we conclude from Step 2 that $G$ is the greatest common factor of $b$ and $r_1$ if and only if $G$ is the greatest common factor of $r_1$ and $r_2$. Thus, the greatest common factor of $a$ and $b$ is the G.C.F. of $r_1$ and $r_2$ and conversely.

By Step 3, $G$ is the G.C.F. of $a$ and $b$ if and only if $G$ is the

## 12.4 Euclidean Algorithm

G.C.F. of $r_2$ and $r_3$. Continuing, we conclude that $G$ is the greatest common factor of $a$ and $b$ if and only if it is the G.C.F. of $r_{n-2}$ and $r_{n-1}$. Since $r_{n-2} = r_{n-1} q_n$, we have $r_{n-1}$ is a factor of $r_{n-2}$; thus, the greatest common factor $G$ of $r_{n-2}$ and $r_{n-1}$ is $r_{n-1}$. Therefore, the G.C.F. of $a$ and $b$ is $r_{n-1}$.

For the positive integers 768 and 426, can we find two integers $u$ and $v$ (not necessarily positive) such that the sum $768u + 426v$ is the greatest common factor 6? The answer is "yes," and the Euclidean algorithm gives us a routine method to find such integers. Notice that the steps on page 216 are equivalent to the following.

Step 1.   $768 = (426)1 + 342$.

Step 2.   $426 = (342)1 + 84$.

Step 3.   $342 = (84)4 + 6$.

$84 = (6)14$.

From Step 3, $6 = 342 - (84)4$; from Step 2, $84 = 426 - (342)1$. Substituting the second equality in the first, we get

$$6 = 342 - [426 - (342)1]4,$$

$$6 = 342 - (426)4 + (342)4,$$

$$6 = (342)5 - (426)4.$$

From Step 1, $342 = 768 - (426)1$; substituting this in the last equality we get

$$6 = [768 - (426)1]5 - (426)4,$$

$$6 = (768)5 - (426)5 - (426)4,$$

$$6 = (768)5 + (426)(-9).$$

Obviously, if we let $u = 5$ and $v = -9$, then we have two integers $u$ and $v$ such that $768u + 426v$ is the greatest common factor of the two numbers.

The Euclidean algorithm can be used to prove for any pair of positive integers $a$ and $b$ with $G$ as their greatest common factor that there exist integers $u$ and $v$ such that

$$G = au + bv.$$

Let us prove this if $a > b$ and if $r_4 = 0$ in the Euclidean algorithm.

Since the number of steps in the Euclidean algorithm is finite, it should be clear that the proof (though tedious) can be generalized to any number of steps. If $r_4 = 0$, recall that

Step 1.    $a = bq_1 + r_1$    where $0 < b$.

Step 2.    $b = r_1 q_2 + r_2$    where $0 < r_2 < r_1$.

Step 3.    $r_1 = r_2 q_3 + r_3$    where $0 < r_3 < r_2$.

Step 4.    $r_2 = r_3 q_4$    where $r_4 = 0$.

Since $G = r_3$, we have from Step 3 that

$$G = r_1 - r_2 q_3.$$

Solving the equality in Step 2 for $r_2$ and substituting in the last equality, we get

$$G = r_1 - [b - r_1 q_2]q_3,$$
$$G = r_1 - bq_3 + r_1 q_2 q_3.$$

Solving the equality in Step 1 for $r_1$ and substituting in the last equality, we get

$$G = (a - bq_1) - bq_3 + (a - bq_1)q_2 q_3,$$
$$G = a - bq_1 - bq_3 + aq_2 q_3 - bq_1 q_2 q_3,$$
$$G = a(1 + q_2 q_3) + b(-q_1 - q_3 - q_1 q_2 q_3).$$

Thus, $u = 1 + q_2 q_3$ and $v = -q_1 - q_3 - q_1 q_2 q_3$ are two integers such that

$$G = au + bv.$$

In our numerical example, $q_1 = 1$, $q_2 = 1$, $q_3 = 4$, and $q_4 = 14$. Notice that $1 + q_1 q_3 = 1 + 4 = 5$ and $-q_1 - q_3 - q_1 q_2 q_3 = -1 - 4 - 4 = -9$ and the previous solution is obtained.

We saw in the last section that it need not be true that if $a|bc$ then $a|b$ or $a|c$. However, we can prove a similar theorem which is quite important and a result of our last theorem.

**Theorem.** If $p$ is a prime number and if $p|bc$, then $p|b$ or $p|c$.

*Proof.* The prime $p$ is a factor of $b$ or it is not. If it is a factor of $b$, then the theorem is true. Thus, suppose $p$ is not a factor of $b$. Since $p$ is a prime and is not a factor of $b$, the greatest common

## 12.4 Euclidean Algorithm

factor of the two numbers is 1. Therefore, there exist integers $u$ and $v$ such that

$$1 = pu + bv.$$

Multiplying both sides by $c$, we obtain

$$c = pcu + bcv.$$

By assumption $p|bc$ so $p|bcv$. Obviously, $p|pcu$. Consequently, $p|(pcu + bcv)$; that is, $p|c$.

### Exercises I

In Exercises 1 through 8, find the G.C.F. of the given set of numbers.
1. $\{689, 474\}$.
2. $\{6,940, 8,964\}$.
3. $\{972, 4,690\}$.
4. $\{3,164, 27,868,484\}$.
5. $\{3,285, 10,731\}$.
6. $\{27,736, 90,856\}$.
7. $\{2,124, 1,548\}$.
8. $\{88,284, 33,880, 8,684\}$.
9. Find the greatest common factor $G$ of 168 and 889, and then find integers $u$ and $v$ such that $G = 889u + 168v$.
10. Find the greatest common factor $G$ of 84 and 194, and then find integers $u$ and $v$ such that $G = 194u + 84v$.

### Exercises II

1. Assume $a|bc$ and assume that $a$ and $b$ have no common factor other than 1. Prove that $a|c$.
2. Assume that $r_5 = 0$ in the Euclidean algorithm and find integers $u$ and $v$ in terms of the successive quotients $q_1, q_2, q_3,$ and $q_4$ such that $G = au + bv$.
3. Show that Exercise 2 applies in finding the greatest common factor $G$ of 889 and 1,946, and then determine a pair of integers $u$ and $v$ such that $G = 1,946u + 889v$.
4. Let $a = b$ and show that there exist integers $u$ and $v$ such

that $G = au + bv$ where $G$ is the greatest common factor of $a$ and $b$.
5. Can we now conclude that if $a$ and $b$ are any positive integers then there exist integers $u$ and $v$ such that $G = au + bv$ where $G$ is the G.C.F. of $a$ and $b$?

## 12.5 Number Congruence

Suppose it is three o'clock. What time would the clock read 138 hours from now? In 12 hours it would be three o'clock. In 24 hours, it would again be three o'clock; in fact, in $12x$ hours where $x$ is a positive integer the time would be three o'clock. After some reflection, it should be obvious that we are first interested in finding the largest multiple of 12 less than or equal to 138. This can be done by the long division algorithm.

$$
\begin{array}{r}
11 \\
12\overline{\smash{)}138} \\
\underline{12} \\
18 \\
\underline{12} \\
6
\end{array}
$$

Therefore, $138 = (12)(11) + 6$ where 11 is the incomplete quotient and 6 is the remainder. Since $138 - (12)(11) = 6$, $(12)(11) = 132$ is the largest multiple of 12 less than or equal to 138. Thus, the clock would read six hours later than three o'clock; that is, it would be nine o'clock.

**Questions.** If it is now five o'clock, what time would it be in each of the following cases? (1) 37 hours from now. (2) 273 hours from now. (3) 3,586 hours from now. (4) 526 hours from now.

From $138 = (12)(11) + 6$, which we obtained above, we see that $138 - 6 = (12)(11)$. Thus, the difference $138 - 6$ has 12 as a factor. For such numbers as 138 and 6 whose difference has 12 as a factor we say that 138 is *congruent* to 6, *modulo* 12. Symbolically, $138 \equiv 6$, mod 12. Congruence for integers is obviously quite different from congruence for plane figures in geometry. The general definition of number congruence is as follows.

## 12.5 Number Congruence

**Definition.** An integer $a$ is said to be *congruent* to an integer $b$, modulo $m$, where $m$ is a *positive integer*, if and only if there is an integer $x$ such that $a - b = mx$. In other words, $a \equiv b$, mod $m$, if and only if $m$ is a factor of $a - b$. The positive integer $m$ is called the *modulus*. (Note that $a$ and $b$ are not necessarily positive integers.)

### Examples

1. $13 \equiv 5$, mod 8, since $8(1) = 13 - 5$.
2. $16 \equiv 2$, mod 7, since $7(2) = 16 - 2$.
3. $26 \equiv 38$, mod 3, since $3(-4) = 26 - 38$.
4. $18 \equiv 0$, mod 6, since $6(3) = 18 - 0$.
5. $11 \equiv 11$, mod 17, since $17(0) = 11 - 11$.
6. $18 \equiv 3$, mod 5.          7. $28 \equiv 36$, mod 4.
8. $15 \equiv 0$, mod 3.           9. $3 \equiv 18$, mod 5.
10. $14 \equiv -4$, mod 9.         11. $-8 \equiv -11$, mod 3.
12. $28 \equiv 28$, mod 3.         13. $38 \equiv 38$, mod 8.

The congruence relation has the same basic properties as those stated earlier for the equals relation. (See Theorem 1 below.) In fact, we shall prove several theorems concerning addition and multiplication for congruences which should look familiar.

**Theorem 1.** Let $m$ be a positive integer. (a) For any integer $a$, $a \equiv a$, mod $m$. (This is called the *reflexive property*.) (2) If $a$ and $b$ are integers such that $a \equiv b$, mod $m$, then $b \equiv a$, mod $m$. (This is called the *symmetric property*.) (3) If $a$, $b$, and $c$ are integers such that $a \equiv b$, mod $m$, and $b \equiv c$, mod $m$, then $a \equiv c$, mod $m$. (This is called the *transitive property*.)

*Proof. Part 1.* For any integer $a$ and a positive integer $m$, there is an integer $x$, namely zero, such that $a - a = mx$. Thus, $a \equiv a$, mod $m$.
*Part 2.* If $a \equiv b$, mod $m$, then from the definition of congruence there exists an integer $x$ such that $a - b = mx$. Thus, $-(a - b) = -mx$, or $b - a = m(-x)$. From the definition of congruence, $b \equiv a$, mod $m$.
*Part 3.* If $a \equiv b$, mod $m$, then there exists an integer $x$ such that $a - b = mx$. If $b \equiv c$, mod $m$, then there exists an integer $y$ such that $b - c = my$. Adding the two equalities,

$$(a - b) + (b - c) = mx + my$$
$$a - c = m(x + y).$$

Since $x + y$ is an integer, $a \equiv c$, mod $m$.

**Theorem 2.** Let $m$ be a positive integer. If $a$, $b$, and $c$ are integers such that $a \equiv b$, mod $m$, then $a + c \equiv b + c$, mod $m$.

*Proof.* If $a \equiv b$, mod $m$, then there is an integer $x$ such that
$$mx = a - b.$$
Thus,
$$mx = (a + c) - (b + c).$$
Therefore, from the definition of the congruence relation,
$$a + c \equiv b + c, \text{ mod } m.$$

**Theorem 3.** Let $m$ be a positive integer. If $a$, $b$, and $c$ are integers such that $a \equiv b$, mod $m$, then $ac \equiv bc$, mod $m$. (Note: If $ac \equiv bc$, mod $m$, then $ca \equiv cb$, mod $m$.)

*Proof.* Left as an exercise.

**Theorem 4.** Let $m$ be a positive integer. If $a \equiv b$, mod $m$, and $c \equiv d$, mod $m$, then $a + c \equiv b + d$, mod $m$.

*Proof.* Since $a \equiv b$, mod $m$, we conclude from Theorem 2 that $a + c \equiv b + c$, mod $m$. Similarly, $c \equiv d$, mod $m$, implies that $b + c \equiv b + d$, mod $m$. By the transitive property of the congruence relation, $a + c \equiv b + d$, mod $m$.

**Theorem 5.** Let $m$ be a positive integer. If $a \equiv b$, mod $m$, and $c \equiv d$, mod $m$, then $ac \equiv bd$, mod $m$.

*Proof.* Left as an exercise.

**Theorem 6.** If $a \equiv b$, mod $m$, then $a^2 \equiv b^2$, mod $m$.

*Proof.* Left as an exercise.

Although $a \equiv b$, mod $m$, implies that $ac \equiv bc$, mod $m$, the

## 12.5 Number Congruence

converse of this theorem is not true. For example, since $24 - 12$ has 6 as a factor, $8 \times 3 \equiv 4 \times 3$, mod 6; however, it is *not* true that $8 \equiv 4$, mod 6. We do have the following theorem which is a restricted cancellation property for multiplication.

**Theorem 7.** If $ac \equiv bc$, mod $m$, and if $m$ and $c$ have no factors in common except 1 and $-1$, then $a \equiv b$, mod $m$.

*Proof.* $ac \equiv bc$, mod $m$, implies that there is an integer $x$ such that

$$mx = ac - bc;$$

thus,

$$mx = c(a - b).$$

Since $m$ is a factor of the product $c(a - b)$ and since $m$ and $c$ are relatively prime, $m$ is a factor of $(a - b)$. Hence,

$$a \equiv b, \text{ mod } m.$$

The concept of number congruence can also be used to prove some of the previously stated divisibility tests for integers. For example, let us prove that the six-digit number $abc,def$ is divisible by 11 if and only if $(b + d + f) - (a + c + e)$ is divisible by 11. First, recall from our positional notation for integers that

$$abc,def = 10^5 a + 10^4 b + 10^3 c + 10^2 d + 10e + f.$$

The following statements are easy to verify from the definition and theorems on congruence.

$$10 \equiv -1, \text{ mod } 11. \quad (A)$$

Squaring (A),

$$10^2 \equiv 1, \text{ mod } 11. \quad (B)$$

Multiplying (A) by (B),

$$10^3 \equiv -1, \text{ mod } 11. \quad (C)$$

Squaring (B),

$$10^4 \equiv 1, \text{ mod } 11. \quad (D)$$

Multiplying (B) by (D),

$$10^5 \equiv -1, \text{ mod } 11.$$

From these equations, it follows that

$$a10^5 \equiv -a, \text{ mod } 11,$$
$$b10^4 \equiv b, \text{ mod } 11,$$
$$c10^3 \equiv -c, \text{ mod } 11,$$
$$d10^2 \equiv d, \text{ mod } 11,$$
$$e10 \equiv -e, \text{ mod } 11,$$
$$f \equiv f, \text{ mod } 11.$$

Adding the last set of congruences,

$$abc,def \equiv (b + d + f) - (a + c + e), \text{ mod } 11.$$

Therefore, $abc,def$ is divisible by 11 if and only if $(b + d + f) - (a + c + e)$ is divisible by 11.

The following theorem exhibits an important relationship between the long division algorithm and the congruence relation. Recall that if $a$ and $m$ are positive integers, then we can find two nonnegative integers $q$ and $r$ such that $a = qm + r$ where $r < m$.

**Theorem 8.** Let $a$, $b$, and $m$ be *positive* integers. $a \equiv b$, mod $m$, if and only if $a$ and $b$ have the same remainder in the long division of each by $m$.

*Proof. Part 1.* Assume $a \equiv b$, mod $m$ and assume $a \geq b$. Thus, $a - b = q_1 m$ where $q_1$ is a nonnegative integer. In the long division of $b$ by $m$, we have

$$b = q_2 m + r \quad \text{where } 0 \leq r < m.$$

Then, since

$$a = b + q_1 m,$$

we have

$$a = (q_2 m + r) + q_1 m,$$
$$a = (q_1 m + q_2 m) + r,$$
$$a = m(q_1 + q_2) + r \quad \text{where } 0 \leq r < m.$$

Hence, $r$ is also the remainder in the long division of $a$ by $m$.
*Part 2.* Assume that the remainders in the long division of each $a$ and $b$ by $m$ is $r$. Thus,

## 12.5 Number Congruence

$$a = q_1 m + r \quad \text{where } 0 \leq r < m$$

and

$$b = q_2 m + r \quad \text{where } 0 \leq r < m.$$

Subtracting the last two equalities,

$$a - b = q_1 m - q_2 m$$

and

$$a - b = m(q_1 - q_2).$$

Since $q_1 - q_2$ is an integer, $m$ is a factor of $a - b$ and $a \equiv b$, mod $m$.

Let us consider the casting-out-nines check in the setting of congruences. Since

$$10 \equiv 1, \text{ mod } 9,$$
$$10^2 \equiv 1, \text{ mod } 9,$$
$$10^3 \equiv 1, \text{ mod } 9,$$
$$10^4 \equiv 1, \text{ mod } 9,$$
$$10^5 \equiv 1, \text{ mod } 9,$$

it follows that

$$a10^5 \equiv a, \text{ mod } 9,$$
$$b10^4 \equiv b, \text{ mod } 9,$$
$$c10^3 \equiv c, \text{ mod } 9,$$
$$d10^2 \equiv d, \text{ mod } 9,$$
$$e10 \equiv e, \text{ mod } 9,$$
$$f \equiv f, \text{ mod } 9.$$

Adding the last set of congruences,

$$abc,def \equiv a + b + c + d + e + f, \text{ mod } 9.$$

We should see not only how the casting-out-nines check gets its name, but the following example should also show us why it works. Consider $322 \times 84$. Now,

$$322 \equiv 3 + 2 + 2, \text{ mod } 9$$

and

$$84 \equiv 8 + 4, \text{ mod } 9.$$

Since $8 + 4 = 12$ and since $12 \equiv 1 + 2$, mod 9,

and
$$322 \equiv 7, \text{ mod } 9,$$
Multiplying,
$$\frac{84 \equiv 3, \text{ mod } 9.}{27{,}048 \equiv 21, \text{ mod } 9.}$$

But
$$27{,}048 \equiv 2 + 7 + 0 + 4 + 8 = 21 \equiv 2 + 1 = 3, \text{ mod } 9$$
and
$$21 \equiv 2 + 1 = 3, \text{ mod } 9.$$

### Exercises I

In each of Exercises 1 through 10, find the nonnegative integer $N$ less than the modulus which makes the statement true.

1. $37 \equiv N$, mod 5.
2. $387 \equiv N$, mod 11.
3. $427 \equiv N$, mod 8.
4. $2{,}374 \equiv N$, mod 3.
5. $243{,}677 \equiv N$, mod 9.
6. $321{,}573 \equiv N$, mod 9.
7. $3N + 7 \equiv 53$, mod 7.
8. $3N + 23 \equiv 11$, mod 7.
9. $4N + 8 \equiv 23$, mod 5.
10. $2N + 32 \equiv 16$, mod 5.
11. Which of the following congruences has a solution in the given set?
    (a) $2x \equiv 3$, mod 3; $\{0, 1, 2\}$.
    (b) $2x \equiv 3$, mod 4; $\{0, 1, 2, 3\}$.
    (c) $2x \equiv 3$, mod 5; $\{0, 1, 2, 3, 4\}$.
    (d) $2x \equiv 3$, mod 6; $\{0, 1, 2, 3, 4, 5\}$.
    (e) $2x \equiv 3$, mod 7; $\{0, 1, 2, 3, 4, 5, 6\}$.
    (f) $2x \equiv 3$, mod 8; $\{0, 1, 2, 3, 4, 5, 6, 7\}$.
    (g) $2x \equiv 3$, mod 9; $\{0, 1, 2, 3, 4, 5, 6, 7, 8\}$.
    (h) $2x \equiv 3$, mod 10; $\{0, 1, 2, 3, 4, 5, 6, 7, 8, 9\}$.
12. Make a conjecture on the basis of your answers in Exercise 11.

### Exercises II

1. Prove Theorem 3.
2. Prove Theorem 5.
3. Prove Theorem 6.

4. In Theorem 8, explain why there is no loss of generality in assuming $a > b$.
5. Let $m$ and $a$ be positive integers. Prove that $m$ is a factor of $a$ if and only if $a \equiv 0$, mod $m$.
6. Prove if $a \equiv b$, mod $m$ and if $n$ is a positive factor of $m$, then $a \equiv b$, mod $n$.
7. Let $x$ be an odd positive integer. Prove that $x^2$ is congruent to 0, 1, or 4, mod 8.
8. Let $S = \{1, 2, 3, 4, 5, 6\}$. If $x$ is a variable on the set $S$, which of the following congruences has a solution? $1x \equiv 1$, mod 7; $2x \equiv 1$, mod 7; $3x \equiv 1$, mod 7; $4x \equiv 1$, mod 7; $5x \equiv 1$, mod 7; $6x \equiv 1$, mod 7.

## 12.6 Other Number Bases

In order to understand better the base-ten positional notation, let us consider the notation that might have developed if man had only eight fingers instead of ten.

If we are going to denote positive integers using a positional system with eight as the base number, we need to "invent" eight numerals to use. It is probably best to take the easy way out and use the eight familiar numerals 0, 1, 2, 3, 4, 5, 6, and 7. (Remember that "8" and "9" are symbols which have no place in our system.) An advantage of this selection is that "3," for example, denotes the number three and we do not have to learn a new symbol for this number. However, there is a definite disadvantage; the symbol "10" represents *eight and not ten*. In fact, to help avoid confusion, we shall usually write "$10_{\text{eight}}$" (read "one-zero, base eight") to emphasize we are using base-eight notation. In base eight, "$236_{\text{eight}}$" represents two sixty-fours (eight eights), three eights, and six units; that is,

$$236_{\text{eight}} = 2(10_{\text{eight}})^2 + 3(10_{\text{eight}}) + 6.$$

To convert $236_{\text{eight}}$ to base-ten notation, we use the fact that

$$236_{\text{eight}} = 2(8)^2 + 3(8) + 6,$$

where on the right side of the equation we are using base-ten notation. Thus,

$$236_{\text{eight}} = 158 \text{ in base ten.}$$

Let us write down the first seventy-two positive integers in base eight so we can observe the pattern and see how easy it would be to write down the first, say, thousand positive integers in this base.

| 1 | $11_{eight}$ | $21_{eight}$ | $31_{eight}$ | $41_{eight}$ | $51_{eight}$ | $61_{eight}$ | $71_{eight}$ | $101_{eight}$ |
|---|---|---|---|---|---|---|---|---|
| 2 | $12_{eight}$ | $22_{eight}$ | $32_{eight}$ | $42_{eight}$ | $52_{eight}$ | $62_{eight}$ | $72_{eight}$ | $102_{eight}$ |
| 3 | $13_{eight}$ | $23_{eight}$ | $33_{eight}$ | $43_{eight}$ | $53_{eight}$ | $63_{eight}$ | $73_{eight}$ | $103_{eight}$ |
| 4 | $14_{eight}$ | $24_{eight}$ | $34_{eight}$ | $44_{eight}$ | $54_{eight}$ | $64_{eight}$ | $74_{eight}$ | $104_{eight}$ |
| 5 | $15_{eight}$ | $25_{eight}$ | $35_{eight}$ | $45_{eight}$ | $55_{eight}$ | $65_{eight}$ | $75_{eight}$ | $105_{eight}$ |
| 6 | $16_{eight}$ | $26_{eight}$ | $36_{eight}$ | $46_{eight}$ | $56_{eight}$ | $66_{eight}$ | $76_{eight}$ | $106_{eight}$ |
| 7 | $17_{eight}$ | $27_{eight}$ | $37_{eight}$ | $47_{eight}$ | $57_{eight}$ | $67_{eight}$ | $77_{eight}$ | $107_{eight}$ |
| $10_{eight}$ | $20_{eight}$ | $30_{eight}$ | $40_{eight}$ | $50_{eight}$ | $60_{eight}$ | $70_{eight}$ | $100_{eight}$ | $110_{eight}$ |

Starting from 6 in column one, if we "count" five more numbers, we arrive at $13_{eight}$. Thus, $6 + 5 = 13_{eight}$. Since $13_{eight}$ is one eight plus three, this, of course, is 11 in base ten. To find the product $6 \times 5$, we could start in the first column and mark off successively six groups of fives and look at the last number in the last group; it is $36_{eight}$. Of course, to obtain the number in base-eight notation, we can make use of the fact that we know that $6 \times 5$ is thirty (three eights and six units). Before continuing, an instructive and worthwhile task would be to complete the following addition and multiplication tables using base-eight notation.

BASE EIGHT

| + | 0 | 1 | 2 | 3 | 4 | 5 | 6 | 7 |
|---|---|---|---|---|---|---|---|---|
| 0 | | | | | | | | |
| 1 | | | | | | | | |
| 2 | | | | | | | | |
| 3 | | | | | | | | |
| 4 | | | | | | | | |
| 5 | | | | | | | | |
| 6 | | | | | | 13 | | |
| 7 | | | | | | | | |

*Addition Table*

## 12.6 Other Number Bases

**BASE EIGHT**

| × | 0 | 1 | 2 | 3 | 4 | 5 | 6 | 7 |
|---|---|---|---|---|---|---|---|---|
| 0 | | | | | | | | |
| 1 | | | | | | | | |
| 2 | | | | | | | | |
| 3 | | | | | | | | |
| 4 | | | | | | | | |
| 5 | | | | | | | | |
| 6 | | | | | 36 | | | |
| 7 | | | | | | | | |

*Multiplication Table*

We see in the addition and multiplication tables for base eight that each table contains sixty-four entries. Similar tables for base ten each contain one hundred entries. One advantage of a smaller base is that there are less number facts to be memorized.

It is not too difficult to prove that the algorithms (techniques) for performing arithmetic calculations with positive integers depend on the positional notation and not on the base number. The student should recall that the development of these algorithms did not depend on the base number, only on the properties of numbers and the positional notation. Thus, the techniques for doing calculations in any base are essentially the same as for base ten; it is only the addition and multiplication tables and the notation which changes. Let us exhibit this fact by working several examples in base eight and in base two. (The student should check each example carefully using the base-eight addition and multiplication tables.)

**Addition**

| Base Eight | | Base Ten |
|---|---|---|
| $13_{eight}$ | ←———→ | 11 |
| $36_{eight}$ | ←———→ | 30 |
| $41_{eight}$ | ←———→ | 33 |
| $\overline{112_{eight}}$ | ←———→ | $\overline{74}$ |

## Multiplication

$$
\begin{array}{r}
236_{eight} \\
25_{eight} \\
\hline
1426_{eight} \\
474\phantom{0}_{eight} \\
\hline
6366_{eight}
\end{array}
\qquad
\begin{array}{c}
\longleftrightarrow \\
\longleftrightarrow \\
\\
\\
\longleftrightarrow
\end{array}
\qquad
\begin{array}{r}
158 \\
21 \\
\hline
158 \\
316\phantom{0} \\
\hline
3318
\end{array}
$$

In base two, we use "0" and "1" as the two numerals. The addition addition and multiplication tables are as follows.

BASE TWO

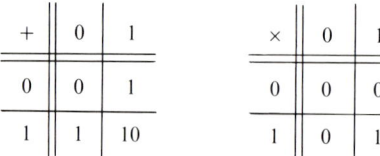

*Addition Table*   *Multiplication Table*

## Examples for addition.

(a) $\quad\begin{array}{r} 1\,0\,0\,1_{two} \\ 1\,0\,1_{two} \\ \hline 1\,1\,1\,0_{two} \end{array}$ 
Check: $\begin{array}{r} 1\,0\,0\,1_{two} = 9 \\ 1\,0\,1_{two} = 5 \\ \hline 1\,1\,1\,0_{two} = 1\,4 \end{array}$

(b) $\quad\begin{array}{r} 1\,1\,1\,1\,0_{two} \\ 1\,1\,0\,1\,1_{two} \\ \hline 1\,1\,1\,0\,0\,1_{two} \end{array}$ 
Check: $\begin{array}{r} 1\,1\,1\,1\,0_{two} = 3\,0 \\ 1\,1\,0\,1\,1_{two} = 2\,7 \\ \hline 1\,1\,1\,0\,0\,1_{two} = 5\,7 \end{array}$

(c) $\quad\begin{array}{r} 1\,0\,1_{two} \\ 1\,1\,1_{two} \\ 1\,1_{two} \\ 1\,0\,0_{two} \\ \hline 1\,0\,0\,1\,1_{two} \end{array}$ 
Check: $\begin{array}{r} 1\,0\,1_{two} = 5 \\ 1\,1\,1_{two} = 7 \\ 1\,1_{two} = 3 \\ 1\,0\,0_{two} = 4 \\ \hline 1\,0\,0\,1\,1_{two} = 1\,9 \end{array}$

## Examples for subtraction.

(a) $\begin{array}{r} 1\,1\,1\,1_{two} \\ 1\,0\,1_{two} \\ \hline 1\,0\,1\,0_{two} \end{array}$
(b) $\begin{array}{r} 1\,0\,1\,1\,0_{two} \\ 1\,0\,1\,1_{two} \\ \hline 1\,0\,1\,1_{two} \end{array}$
(c) $\begin{array}{r} 1\,0\,1\,1\,0\,1_{two} \\ 1\,0\,1\,1\,0_{two} \\ \hline 1\,0\,1\,1\,1_{two} \end{array}$

## 12.6 Other Number Bases

Examples for multiplication.

(a)
$$\begin{array}{r} 1111_{two} \\ 101_{two} \\ \hline 1111 \\ 1111\phantom{00} \\ \hline 1001011_{two} \end{array}$$

(b)
$$\begin{array}{r} 10110_{two} \\ 1011_{two} \\ \hline 10110 \\ 10110\phantom{0} \\ 10110\phantom{00} \\ \hline 11110010_{two} \end{array}$$

(c)
$$\begin{array}{r} 11011_{two} \\ 11000_{two} \\ \hline 11011000 \\ 11011\phantom{000} \\ \hline 101000100 0_{two} \end{array}$$

(d)
$$\begin{array}{r} 10111 0_{two} \\ 1001_{two} \\ \hline 101110 \\ 101110\phantom{000} \\ \hline 110011110_{two} \end{array}$$

Examples for long division.

(a)
$$\begin{array}{r} 11_{two} = \text{quotient} \\ 101_{two} \overline{\smash{)}1111_{two}} \\ \underline{101\phantom{0}} \\ 101 \\ \underline{101} \\ 0 = \text{remainder} \end{array}$$

(b)
$$\begin{array}{r} 11100_{two} = \text{quotient} \\ 1101_{two} \overline{\smash{)}101110111_{two}} \\ \underline{1101\phantom{00000}} \\ 10100 \\ \underline{1101\phantom{0}} \\ 1111 \\ \underline{1101} \\ 1011_{two} = \text{remainder} \end{array}$$

There is a simple and routine procedure to change numbers from base-ten notation to base-two notation. To change, say, 46 from base ten to base two, we proceed as follows.

Step 1. $2 \underline{|\ 46}$

Step 2. $2 \underline{|\ 23} \quad r = 0$

Step 3. $2 \underline{|\ 11} \quad r = 1$

Step 4. $2 \underline{|\ 5} \quad r = 1$

Step 5.   2 | 2    $r = 1$
Step 6.        1    $r = 0$

*Explanation of Procedure.* Using long division (sometimes called short division but it would be more appropriate to say that this is a short method for finding the incomplete quotient and the remainder), divide 46 by 2 and then successively divide each (incomplete) quotient by 2 until a quotient of 1 is obtained; record the remainders after each division. Then, start with the final quotient 1 and list successively after it the remainders in the opposite order from which they were obtained; that is, 1 0 1 1 1 0. We assert that $46 = 101110_{two}$.

To verify this technique, let us consider the steps in the long division process.

Step 1.        $46 = 2(23)$.

Step 2.        $23 = 2(11) + 1$.

Hence,

$$46 = 2[2(11) + 1]$$

and

$$46 = 2^2(11) + 2.$$

Step 3.        $11 = 2(5) + 1$.

Hence,

$$46 = 2^2[2(5) + 1] + 2$$

and

$$46 = 2^3(5) + 2^2 + 2.$$

Step 4.        $5 = 2(2) + 1$.

Hence,

$$46 = 2^3[2(2) + 1] + 2^2 + 2$$

and

$$46 = 2^4(2) + 2^3 + 2^2 + 2.$$

Step 5.        $46 = 2^5 + 2^3 + 2^2 + 2.$

In other words,

## 12.6 Other Number Bases

$$46 = 2^5 + 0(2^4) + 2^3 + 2^2 + 2 + 0$$

which states in positional notation for base two that

$$46 = 1\ 0\ 1\ 1\ 1\ 0_{\text{two}}.$$

Let us review our halve-double-sum method for multiplication that we discussed on page 76. There we found the product $46 \times 128$ as follows.

```
    A           B
   -46-       -128-
    23         256  ←──┐
    11         512  ←──┤
     5        1024  ←──┤ ←── Add
    -2-      -2048-    │
     1        4096  ←──┘
             ─────
             5888   Sum
```

Since $46 = 1\ 0\ 1\ 1\ 1\ 0_{\text{two}} = 2^5 + 0(2^4) + 2^3 + 2^2 + 2 + 0$, it follows that

$$46 \times 218 = (2^5 + 2^3 + 2^2 + 2) \times 128$$
$$= 2^5(128) + 2^3(128) + 2^2(128) + 2(128)$$
$$= 4096 + 1024 + 512 + 256$$
$$= 5888.$$

This example should indicate how the halve-double-sum procedure could be verified.

### Exercises I

1. Complete the base-eight addition and multiplication tables on pages 228 and 229.
2. (a) State rules for carrying when adding in base two.
   (b) State rules for borrowing when subtracting in base two.
3. Check each subtraction problem in the worked examples in base two by using base-ten notation.
4. Check each multiplication problem in the worked examples in base two by using base-ten notation.
5. Check each long division problem in the worked examples by finding the product of the divisor and incomplete

quotient and then adding the remainder to get the dividend. (Do your calculations in base two.)

6. Find the following sums by using base two notation. Check your results by using base-ten notation.
   (a) $1101_{two}$
       $\underline{\phantom{1}100_{two}}$
   (b) $1100_{two}$
       $1111_{two}$
       $\underline{1011_{two}}$
   (c) $11111_{two}$
       $1001_{two}$
       $\underline{1111_{two}}$

7. Find the differences by using base-two notation. Check your results by using base-ten notation.
   (a) $1101_{two}$
       $\underline{\phantom{110}11_{two}}$
   (c) $101100_{two}$
       $\underline{10011_{two}}$
   (b) $11010_{two}$
       $\underline{\phantom{1}1111_{two}}$

8. Find the products by using base-two notation. Check your results by using base-ten notation.
   (a) $1101_{two}$
       $\underline{\phantom{110}11_{two}}$
   (c) $101101_{two}$
       $\underline{11100_{two}}$
   (b) $11010_{two}$
       $\underline{1010_{two}}$

9. Use an algorithm similar to the one discussed on page 231 to change the following numbers from base-ten to base-eight notation. (a) 7,894. (b) 1,354. (c) 243,567. (d) 24,640.

10. Find the sums using base-eight notation. Check your results by using base-ten notation.
    (a) $476_{eight}$
        $233_{eight}$
        $421_{eight}$
        $\underline{765_{eight}}$
    (b) $254_{eight}$
        $377_{eight}$
        $35_{eight}$
        $\underline{62_{eight}}$
    (c) $2345_{eight}$
        $1000_{eight}$
        $4010_{eight}$
        $\underline{7650_{eight}}$

11. Find the differences using base-eight notation. Check your results by using base-ten notation.
    (a) $6744_{eight}$
        $\underline{3522_{eight}}$
    (b) $7354_{eight}$
        $\underline{2716_{eight}}$
    (c) $7245_{eight}$
        $\underline{4766_{eight}}$

12. Find the following products by using base-eight notation. Check your results by using base-ten notation.
    (a) $134_{eight}$
    $\phantom{(a)\ }21_{eight}$
    (b) $354_{eight}$
    $\phantom{(b)\ }316_{eight}$
    (c) $2361_{eight}$
    $\phantom{(c)\ }523_{eight}$

13. Use an algorithm similar to the one discussed on page 231 to change each of the following numbers from base-ten to base-six notation. (a) 8,964. (b) 2,345. (c) 467,960. (d) 3,600.

## Exercises II

1. Let $abc,def$ be a six-digit number in base eight where 0, 1, 2, 3, 4, 5, 6, and 7 are the numerals used. State (and prove if possible) your own divisibility tests for 2, 7, and $10_{eight}$.
2. Prove in base three that a six-digit number is divisible by 2 if and only if the sum of the digits is divisible by 2.

## 12.7 Infinite Decimals

Recall our study of geometric sequences in Section 11.8 and consider the following geometric series:

$$1 + \frac{1}{2} + \frac{1}{2^2} + \frac{1}{2^3} + \frac{1}{2^4} + \cdots + \frac{1}{2^{n-1}}.$$

This is a geometric series with $n$ terms where $a = 1$ and $r = \frac{1}{2}$. Thus, the sum is given by

$$S_n = \frac{a - ar^n}{1 - r}$$

$$= \frac{1 - (\frac{1}{2})^n}{1 - \frac{1}{2}}$$

$$= \frac{1 - (\frac{1}{2})^n}{\frac{1}{2}}.$$

Therefore,

$$S_n = 2 - (\tfrac{1}{2})^{n-1}.$$

By choosing $n$ large enough, we can make $(\frac{1}{2})^{n-1}$ as small as we wish and, hence, make $[2 - (\frac{1}{2})^{n-1}]$ as close to 2 as we desire. In fact, we can find a positive integer $N$ such that $N$ terms of this geometric series will differ from 2 by less than, say, one-millionth. Furthermore, adding more terms will just make the sum closer to 2.

Since the sum of the geometric series can be made as close to 2 as we wish by choosing $N$ large enough and since adding more terms just makes the sum closer to 2, we say that 2 is the *sum* of the *infinite geometric series* denoted by

$$1 + \frac{1}{2} + \frac{1}{2^2} + \frac{1}{2^3} + \cdots + \frac{1}{2^{n-1}} + \cdots.$$

We have introduced a new kind of "sum." A careful theoretical approach would make it necessary to prove theorems which would show under what circumstances "infinite sums" have the same properties as "finite sums," such as the associative and commutative properties of addition. However, since our study of infinite sums is to be quite restricted we shall simply state that the infinite geometric sums which we shall study do have the familiar additive properties.

There is a large branch of mathematical analysis dealing with infinite sums and related topics. We shall discuss only a small part of this important topic, but it does include the ideas basic to an understanding of infinite decimals.

For any geometric series where $r \neq 1$, we have that

$$S_n = \frac{a}{1-r} - \frac{ar^n}{1-r}.$$

If $r$ is greater than $-1$ and less than 1, we can make $r^n$ as close to zero as we wish by choosing $n$ large. Hence, $\frac{ar^n}{1-r}$ can be made arbitrarily small and $S_n$ can be made as close to $\frac{a}{1-r}$ as we choose. Therefore, we make the following definition.

**Definition.** The *sum*, $S$, of the infinite geometric series

$$a + ar + ar^2 + ar^3 + \cdots + ar^{n-1} + \cdots,$$

where $r > -1$ and $r < 1$ is

$$\frac{a}{1-r}.$$

We now have the tools necessary to discuss the meaning of the notation "$0.333\overline{3}\ldots$". The ellipses indicate that there is no last

## 12.7 Infinite Decimals

digit and the overbar above the numeral indicates that the symbol "3" continues to repeat. We refer to the notation as an *infinite repeating decimal*, or *repeating decimal*.

The infinite decimal notation is a natural extension of the terminating decimal notation discussed in Chapter 8. The infinite decimal notation "0.333̄..." represents the infinite sum

$$\frac{3}{10} + \frac{3}{10^2} + \frac{3}{10^3} + \frac{3}{10^4} + \cdots.$$

This infinite sum is an infinite geometric series where $a = \frac{3}{10}$ and $r = \frac{1}{10}$. Hence, the sum $S$ of this series, by our formula, is

$$S = \frac{\frac{3}{10}}{1 - \frac{1}{10}} = \frac{\frac{3}{10}}{\frac{9}{10}} = \frac{3}{9} = \frac{1}{3}.$$

Hence,

$$\frac{1}{3} = 0.333\bar{3}\ldots.$$

Consider the infinite repeating decimal $0.6767\overline{67}\ldots$. This decimal would represent the sum of the infinite geometric series

$$\frac{67}{10^2} + \frac{67}{10^4} + \frac{67}{10^6} + \frac{67}{10^8} + \cdots$$

where $a = \frac{67}{10^2}$ and $r = \frac{1}{10^2}$. The sum is the rational number

$$\frac{\frac{67}{10^2}}{1 - \frac{1}{10^2}} = \frac{67}{99};$$

thus, $\frac{67}{99} = 0.6767\overline{67}\ldots$.

The repeating decimal "0.333̄..." is said to have *period one*, and the repeating decimal "0.6767\overline{67}..." is said to have *period two*. The *period* of a repeating decimal is the number of digits in the shortest repeating part of the decimal notation.

### Examples

1. $0.666\bar{6}\ldots$ is an infinite repeating decimal with period one.
2. $0.128128\overline{128}\ldots$ is an infinite repeating decimal with period three.
3. $14.2323\overline{23}\ldots$ is an infinite repeating decimal with period two.

4. $0.12473838\overline{38}\ldots$ is an infinite repeating decimal with period two.

To find the rational number represented by the infinite repeating decimal "$0.23444\overline{4}\ldots$", we proceed as follows.

$$0.23444\overline{4}\ldots = \frac{23}{10^2} + \left[\frac{4}{10^3} + \frac{4}{10^4} + \frac{4}{10^5} + \cdots\right]$$

$$= \frac{23}{10^2} + \frac{\frac{4}{10}^3}{1 - \frac{1}{10}}$$

$$= \frac{23}{100} + \frac{4}{1000 - 100}$$

$$= \frac{207 + 4}{900}$$

$$= \frac{211}{900}.$$

Every infinite repeating decimal can be expressed as an infinite geometric series or as the sum of a rational number and an infinite geometric series. The first term $a$ in the geometric series is a rational number and the common ratio $r$ is also a rational number; in fact, the ratio is $\frac{1}{10^p}$ where $p$ is the period of the repeating decimal. Since the ratio is positive and less than 1, the sum of the infinite geometric series is the rational number

$$\frac{a}{1 - \frac{1}{10^p}}.$$

Therefore, *every repeating infinite decimal represents a rational number.*

The following technique can also be used to find a fractional representation for the rational number represented by a given repeating decimal.

To find a fractional representation for $0.6464\overline{64}\ldots$, let $x = 0.6464646\overline{64}\ldots$. Multiplying by 100, we obtain

$$100x = 64.64646\overline{64}\ldots$$
$$x = 0.64646\overline{64}\ldots.$$

Subtracting, $\quad 99x = 64.$

Thus, $\quad x = \dfrac{64}{99}.$

## 12.7 Infinite Decimals

In this technique, we use the assumed basic properties for infinite sums. In the first step, we assume that

$$100\left(\frac{64}{10^2} + \frac{64}{10^4} + \frac{64}{10^6} + \cdots\right) = 64 + \frac{64}{10^2} + \frac{64}{10^4} + \cdots;$$

that is, we use an "infinite" distributive property. In the second step, we assume that

$$\left(64 + \frac{64}{10^2} + \frac{64}{10^4} + \frac{64}{10^6} + \cdots\right) - \left(\frac{64}{10^2} + \frac{64}{10^4} + \frac{64}{10^6} + \cdots\right) = 64;$$

that is, we use "infinite" associative and commutative properties.

To find a fractional representation for $2.134\overline{34}\ldots$, let

$$\begin{aligned} x &= 2.134343434\ldots \\ 10x &= 21.34343434\ldots \\ 1{,}000x &= 2{,}134.343434\ldots \\ 1{,}000x - 10x &= 2{,}134 - 21 \\ 990x &= 2{,}113 \\ x &= \frac{2{,}113}{990}. \end{aligned}$$

It is also true that every rational number can be represented by an infinite repeating decimal; the repeating decimal representation of a rational number given in fractional notation, such as "$\frac{3}{7}$," is found by the long division algorithm.

$$\begin{array}{r} 0.428571\ldots \\ 7\,\overline{\smash{\big)}\,3.000000} \\ \underline{2\,8}\phantom{0000} \\ 20\phantom{000} \\ \underline{14}\phantom{000} \\ 60\phantom{00} \\ \underline{56}\phantom{00} \\ 40\phantom{0} \\ \underline{35}\phantom{0} \\ 50\phantom{0} \\ \underline{49}\phantom{0} \\ 10 \\ \underline{7} \\ 30. \end{array}$$

Since the next step in this process is the same as the first step, we have $\frac{3}{7} = 0.428571\overline{428571}\ldots$.

For any given rational number $\frac{p}{q}$, the repeating decimal representation can be found by the long division algorithm. In the algorithm, we know that the remainder is always a whole number less than the divisor; thus, it should be evident that the period of the repeating decimal representation of $\frac{p}{q}$ must be less than the denominator $q$.

We now have that *every rational number may be represented by an infinite repeating decimal and every infinite repeating decimal represents a rational number.*

## Exercises I

1. Express the sum of the following infinite geometric series as a fraction: $\frac{1}{7} + \frac{1}{14} + \frac{1}{28} + \frac{1}{56} + \cdots$.
2. Express the sum of the following infinite geometric series as a fraction: $1 + \frac{1}{3} + \frac{1}{9} + \frac{1}{27} + \cdots$.

In Exercises 3 through 6, use the formula for the sum of an infinite geometric series to find a fractional representation for the infinite repeating decimal.

3. $0.218218\overline{218}\ldots$
4. $0.423423\overline{423}\ldots$
5. $0.99999\overline{9}\ldots$
6. $3.237777\overline{7}\ldots$

Use the second technique discussed in the text to find a fractional representation for the infinite repeating decimals in Exercises 7 through 12.

7. $0.212121\overline{21}\ldots$
8. $0.100100\overline{100}\ldots$
9. $0.315315\overline{315}\ldots$
10. $0.239898\overline{98}\ldots$
11. $36.141414\overline{14}\ldots$
12. $0.2763843\overline{3843}\ldots$

In Exercises 13 through 18, express each rational number as an infinite repeating decimal.

13. $\frac{5}{7}$.
14. $\frac{2}{9}$.
15. $\frac{3}{11}$.
16. $\frac{4}{17}$.
17. $\frac{3}{8}$.
18. $\frac{11}{8}$.

## 12.8 Irrational Numbers

We have discussed a method to associate each rational number with a point on the number line. However, there exist points on the

## 12.8 Irrational Numbers

number line that do not correspond to any rational number. For example, we can obtain one in the following manner. Construct an angle of 45° at the origin of the number line as in Figure 1. Let $A$ be the point on the side of the angle such that the length of $OA$ is 1. Construct a perpendicular line to $OA$ at $A$; let $B$ be the point of intersection of this line and the number line. Since triangle $OAB$ is isosceles, the length of $AB$ is 1.

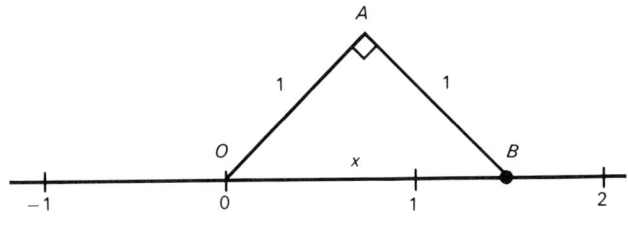

Figure 1. $x^2 = 2$.

By the Pythagorean Theorem (in a right triangle, the sum of the squares of the lengths of the sides is equal to the square of the length of the hypotenuse), we conclude that $(OB)^2 = 1^2 + 1^2 = 2$. Therefore, a number used to denote the length of $OB$ would be the number whose square is 2. We shall now prove that $B$ is not a rational point on the number line; that is, we prove that no rational number exists such that its square is 2.

**Theorem.** There is no rational number $\frac{p}{q}$ such that $(\frac{p}{q})^2 = 2$.

*Proof.* Assume there is a rational number $\frac{p}{q}$ where $p$ and $q$ have no common factors except 1 and $-1$ such that $(\frac{p}{q})^2 = 2$. (There is no loss in generality by assuming that $p$ and $q$ have no common factors other than 1 and $-1$ since every rational can be expressed by such a fraction.) If

$$\left(\frac{p}{q}\right)^2 = 2,$$

then

$$\frac{p^2}{q^2} = 2;$$

hence,

$$p^2 = 2q^2. \tag{1}$$

Since $q$ is an integer, $q^2$ is an integer and $2q^2$ is an *even* integer. Since $p^2 = 2q^2$, $p^2$ is even. Thus, since the square of every odd integer is odd, $p$ must be an even integer. Since $p$ is even it can be expressed as $p = 2t$ where $t$ is some positive integer. Substituting in Equation (1),

$$(2t)^2 = 2q^2,$$
$$4t^2 = 2q^2,$$
$$2t^2 = q^2.$$

Since $2t^2$ is an even integer, by an argument similar to the one above, we conclude that $q$ is an even integer.

Therefore, we have proved that $p$ and $q$ have a common factor of 2, a contradiction to our original assumption that they had no common factors except 1 and $-1$. Hence, our assumption that there was a rational number whose square is 2 is false, and our theorem is proved.

The existence of nonrational points on the number line, or the lack of closure with respect to the operation of square root, motivates the next extension of the number system. The "new" numbers we introduce are called *irrational numbers*; points on the number line which are not associated with rational numbers are called *irrational points*. The notation "$\sqrt{2}$" is used to denote the positive irrational number whose square is 2.

As the reader may check, the square of each rational number after the first in the sequence 1.4, 1.41, 1.414, 1.4142, 1.41421, 1.414214 differs from 2 by less than what the square of the preceding term in the sequence differs from 2. In fact, one can find a rational approximation of 2 to any degree of accuracy.

It can be proved that for any integer which is not a perfect square no rational number exists such that its square is the given integer; it can be proved that for any integer which is not a perfect cube no rational number exists such that its cube is the given integer; etc. Thus, the set of irrational numbers is an infinite set. Numbers such as $\sqrt{2}$, $\sqrt[5]{3}$, $\sqrt[7]{5}$, $\sqrt[3]{19}$, and $\sqrt[15]{11}$ are not the only "types" of irrational numbers; for example, the number denoted by "$\pi$" is an irrational number.

It can be proved that every irrational number can be represented by an infinite decimal; for example, $\sqrt{2} = 1.414214\ldots$, $\pi = 3.141592\ldots$, $\sqrt{3} = 1.73205\ldots$, etc. We know that the decimal

## 12.8 Irrational Numbers

representations of irrationals cannot be a repeating decimal since repeating decimals represent rational numbers. The infinite series associated with a nonrepeating decimal is not geometric; therefore, further, and more complicated, concepts would be necessary for an accurate discussion of the infinite decimal notation for irrational numbers.

The set consisting of the rationals and the irrationals is called the *real numbers*; every real number can be denoted by an infinite decimal and every infinite decimal represents a real number. As the reader might expect, it can be proved that the set of real numbers satisfies all of the basic properties for the set of rationals.

The set of real numbers is assumed to be in one-to-one correspondence with the points on the number line. Geometrically speaking, we have filled up the "gaps" on the number line with the irrational numbers.

In algebra, the basic operations are addition, subtraction, multiplication, division, and taking roots; these are called the *algebraic operations*. We have found that it is necessary to introduce the irrational numbers so that there will exist a number $x$ such that $x^2 = 2$; to have a number $x$ such that $x^2 = -4$, it is necessary to introduce the *complex numbers*.

The introduction of the complex numbers is the final extension of our number system. As we have found, the set of rational numbers is closed with respect to all the operations of arithmetic (rational operations), except division by zero; it can be proved that the set of complex numbers is closed with respect to all the algebraic operations except division by zero.

In arithmetic, we are interested in the rational operations; a detailed discussion of the real numbers or the complex numbers would not be relevant to this book. For a detailed and rigorous construction of the various number systems, the reader should study, for example, Bevan K Youse, *The Number System* (Belmont, Calif.: Dickenson Publishing Company, Inc., 1965).

### Exercises I

Prove there is no rational number whose square is 3.

# MATHEMATICAL TERMS AND BASIC PROPERTIES

**Addend.** One of the numbers to be added when performing the operation of addition.

**Additive inverse.** For any number $x$, the additive inverse of $x$ is the number $y$ such that $x + y = 0$.

**Algorithm.** Routine procedure or technique for performing an operation.

**Area of a circle.** Area $= \pi r^2$ where $r$ is the radius and $\pi = 3.1416$, correct to four decimal places.

**Area of a parallelogram.** Area $= bh$ where $b$ is the length of the base (one side) and $h$ is the perpendicular distance from the base to the opposite side.

**Area of a rectangle.** Area $= lw$ where $l$ is the length and $w$ is the width of the rectangle.

**Area of a square.** Area $= x^2$ where $x$ is the length of each side.

**Area of a trapezoid.** Area $= \frac{1}{2}(b_1 + b_2)h$ where $b_1$ and $b_2$ are the lengths of the parallel sides and $h$ is the perpendicular distance from one of these sides to the other.

**Arithmetic sequence.** A sequence of numbers where each term after the first is the sum of the preceding term and a fixed constant called the common difference.

**Associative property for the intersection of sets.** For sets $A$, $B$, and $C$, $A \cap (B \cap C) = (A \cap B) \cap C$.

**Associative property for the union of sets.** For sets $A$, $B$, and $C$, $A \cup (B \cup C) = (A \cup B) \cup C$.

**Associative property of addition.** For numbers $a$, $b$, and $c$, $a + (b + c) = (a + b) + c$.

**Associative property of multiplication.** For numbers $a$, $b$, and $c$, $a(bc) = (ab)c$.

**Base number.** The number of objects put into basic groupings for the purposes of counting. Also, the number of basic symbols in a positional notation system for the counting numbers.

**Binary operation.** A method for associating a unique element with a pair of given elements.

**Cardinal number.** A number which designates the size of a set.

**Casting-out-nines.** A check for addition, subtraction, and multiplication of natural numbers.

**Circle.** A plane figure consisting of all points in a plane equidistant from a fixed point in the plane called the center.

**Closure property of addition.** A set $S$ is closed with respect to addition if for any numbers $a$ and $b$ in $S$ the sum $a + b$ is a unique number in the set.

**Closure property of multiplication.** A set $S$ is closed with respect to multiplication if for any numbers $a$ and $b$ in $S$ the product $ab$ is a unique number in the set.

**Common difference.** In an arithmetic sequence, it is the fixed difference obtained by subtracting from any term, after the first, the preceding term.

**Common ratio.** In a geometric sequence, it is the fixed quotient obtained by dividing any term, after the first, by the preceding term.

**Composite number.** A positive integer which has more than two different positive factors.

**Commutative property for the intersection of sets.** For sets $A$ and $B$, $A \cap B = B \cap A$.

**Commutative property for the union of sets.** For sets $A$ and $B$, $A \cup B = B \cup A$.

**Commutative property of addition.** For numbers $a$ and $b$, $a + b = b + a$.

**Commutative property of multiplication.** For numbers $a$ and $b$, $ab = ba$.

**Complement of a set $A$ relative to a set $S$.** Where $A$ is a subset of $S$, it is the set consisting of all elements in $S$ which are not in $A$.

**Decimal fraction.** A simple fraction in which the denominator is an integral power of ten.

**Decimal point.** A period used as separatrix between the units' and tenths' positions in the positional notation system for numbers.

**Decimal system.** The positional system of notation for numbers using ten as the base number.

# Mathematical Terms and Basic Properties

**Denominator.** In the fractional notation $\frac{a}{b}$ for rational numbers, it is the number $b$.

**Digits.** Another name for the basic one-character symbols used to denote numbers.

**Disjoint sets.** Sets which have no elements in common.

**Distributive property.** For numbers $a$, $b$, and $c$, $a(b + c) = ab + ac$.

**Dividend.** In the incomplete division of $a$ by $b$, it is the number $a$.

**Division.** A rational operation defined in terms of multiplication.

**Divisor.** In the incomplete division of $a$ by $b$, it is the number $b$.

**Duodecimal system.** The positional system of notation for numbers using twelve as the base number.

**Euclidean algorithm.** A technique for finding the greatest factor in the set of common factors of two given numbers.

**Euler's phi-function.** A formula which gives the number of positive integers which are less than a positive integer and relatively prime to it.

**Factor.** A natural number $a$ is a factor of a natural number $b$ if and only if there is a natural number $x$ such that $ax = b$.

**Factor tree.** A schematic technique to express a composite number as a product of prime numbers.

**Finite decimal.** An $n$-place terminating decimal.

**Fraction.** The symbol "$\frac{a}{b}$" used to denote rational numbers.

**Fundamental Theorem of Arithmetic.** Every composite number can be expressed as the product of prime numbers in one and only one way.

**Geometric sequence.** A sequence of numbers where each term after the first is the product of the preceding term and a fixed (nonzero) constant called the common ratio.

**Greater-than.** A number $a$ is greater than a number $b$ if and only if there is a positive number $x$ such that $a = b + x$; that is, if and only if $b$ is less than $a$.

**Greatest common factor.** For two or more numbers, it is the greatest number in the set of common factors of the given numbers.

**Halve-double-sum algorithm.** A technique to find the product of two natural numbers.

**Hexagon.** A six-sided polygon.

**Hindu–Arabic numerals.** The symbols 0, 1, 2, 3, 4, 5, 6, 7, 8, and 9.

**Improper fraction.** A fraction where the numerator and denominator are positive integers and the numerator is greater than or equal to the denominator. Thus, it represents a rational number greater than or equal to one.

**Incomplete division.** The operation on two positive integers $a$ and $b$ which determines integers $q$ and $r$ such that $a = bq + r$ where $q \geq 0$ and $0 \leq r < b$.

**Incomplete quotient.** In the incomplete division of $a$ by $b$ it is the integer $q$ where $a = bq + r$, $q \geq 0$ and $0 \leq r < b$.

**Inequalities.** Statements of the form $a < b$, $a > b$, $a \leq b$, or $a \geq b$.

**Intersection of two sets.** The set consisting of the elements common to both sets.

**Least common denominator.** For a given set of fractions, it is the least common multiple of the denominators.

**Least common multiple.** For a given set of two or more numbers, it is the least number which is a multiple of each of the numbers in the set.

**Less-than.** A number $a$ is less than a number $b$ if and only if there is a positive number $x$ such that $a + x = b$.

**Minuend.** In the difference $a - b$, it is the number $a$.

**Mixed numeral notation.** A notation for rational numbers of the form $q\frac{r}{b}$ representing the sum of a positive integer $q$ and a positive rational number $\frac{r}{b}$ which is less than one.

**Multiple.** A number $b$ is a multiple of a number $a$ if and only if $a$ is a factor of $b$.

**Multiplicative inverse.** For any number $x$ different from zero, it is the number $y$ such that $xy = 1$.

**Natural number.** A counting number such as 1, 2, 3, 4, etc.

**Numerals.** The basic symbols used to denote numbers.

**Numerator.** In the fractional notation $\frac{a}{b}$ for rational numbers, it is the number $a$.

**One-to-one correspondence.** A pairwise matching between all the elements of two sets.

**Pentagon.** A polygon with five sides.

**Perimeter.** For a polygon, it is the sum of the lengths of the sides. For a closed curve, it is the length of the curve.

*Mathematical Terms and Basic Properties* 249

**Polygon.** A simple closed figure in the plane consisting of line segments.
**Polyhedron.** A three-dimensional solid in which the faces are polygons.
**Prime number.** A natural number different from the number one and which has only itself and one as factors.
**Proper fraction.** A simple fraction representing a positive rational number which is less than one.

**Quadrilateral.** A polygon with four sides.

**Rational number.** A number which can be expressed as the ratio of two integers.
**Rational operations.** The operations of addition, subtraction, multiplication, and division.
**Reciprocal of a fraction.** For any fraction $\frac{a}{b}$ denoting a number different from zero, it is the inverted fraction $\frac{b}{a}$.
**Reflexive property of equality.** For any number $a$, $a = a$.
**Regular polygon.** A polygon with all sides equal and all interior angles equal.
**Relatively prime.** Two natural numbers are relatively prime if and only if they have no common positive factor except one.
**Remainder.** In the incomplete division of $a$ by $b$, it is the number $r$ where $a = bq + r$, $q \geq 0$, and $0 \leq r < b$.
**Roman numerals.** The symbols I, V, X, L, C, D, and M used to denote numbers.

**Sieve of Eratosthenes.** An algorithm for identifying prime numbers.
**Sphere.** A three-dimensional figure consisting of all points equidistant from a point within called the center.
**Square.** A four-sided regular polygon.
**Subtraction.** A rational operation defined in terms of addition.
**Subtrahend.** In the difference $a - b$, it is the number $b$.
**Symmetric property of equality.** If $a = b$ then $b = a$.

**Transitive property of equality.** If $a = b$ and $b = c$ then $a = c$.
**Transitive property of the less-than relation.** If $a < b$ and $b < c$ then $a < c$.
**Trapezoid.** A four-sided polygon with exactly one pair of parallel sides.
**Triangle.** A polygon with three sides.

**Trichotomy property.** For numbers $a$ and $b$, one and only one of the following is true: $a < b$, $a = b$, or $a > b$.

**Union of two sets.** The set consisting of all elements which are in one or the other (or both) of the given sets.

**Whole number.** A natural number or zero.

**Volume of a sphere.** $V = \frac{4}{3}\pi r^3$ where $r$ is the radius of the sphere and $\pi = 3.1416$, correct to four decimal places.

# ANSWERS TO ODD-NUMBERED EXERCISE PROBLEMS

## Section 1.1, page 6

**Exercises I**

1. The notations for numbers, the basic operations performed on numbers, the techniques for performing these operations, and the use and applications of numbers in everyday pursuits.
3. Ten.
5. It gives the position (ranking) of an element in a set where the elements have been linearly arranged (ordered).
7. The natural numbers and zero.
9. Cardinal; cardinal.
11. Cardinal; ordinal.

**Exercises II**

1. Pair, duet, twin, twice, couple, couplet, brace, double, duplex, duplicate.
3. Sixty seconds in a minute. Sixty minutes in an hour. Sixty seconds in a degree. Sixty groups of sixty degrees in a circle.
5. dodoty-do, dodoty-to, dodoty-fo, todoty, todoty-un, todoty-do, todoty-to, todoty-fo, fodoty, fodoty-un, fodoty-do, fodoty-to, fodoty-fo.
7. (a) Thirty-five (b) Twenty-nine.
9. (a) Sixty. (b) Sixty-two.
11. (a) Seventy-eight. (b) One hundred eight.

## Section 1.2, page 10

**Exercises I**

1. (a) 0, 1, 2, 3, 4, 5, 6, 7, 8, 9.
   (b) I, V, X, L, C, D, M.
3. (a) One hundred twenty-six thousand, eight hundred twenty.
   (b) Three million, one hundred six thousand, four hundred six.
   (c) Twelve million, forty-one.
5. (a) Forty-four.
   (b) One thousand, two hundred sixty-five.
   (c) One thousand, nine hundred twenty-two.
7. (a) 48; XLVIII.        (b) 131; CXXXI.
9. (a) 94; XCIV.          (b) 1492; MCDXCII.

**Exercises II**

1. "Digit" comes from the Latin *digitus* meaning finger.
3. ᙆᙆ, ᙆƐ, ᙆᏢ, ƐO, Ɛ⇂, Ɛᙆ, ƐƐ, ƐᏢ, ᏢO, Ꮰ⇂, Ꮲᙆ, ƐᏢ, ᏢᏢ, ⇂OO, ⇂O⇂.
5. (a) ⇂ᙆO.    (b) ⇂OᏢ.
7. (a) ᙆᙆO.    (b) ᙆᙆƐ.
9. (a) ƐOƐ.    (b) ᏢƐO.
11. Three thousand, six hundred. Eighty-six thousand, four hundred. Forty-one million, five hundred thirty-six thousand.
13. Three hundred seventy-three billion.
15. Ninety-three million.
17. Three billion, six hundred seventy-five million, two hundred seventy thousand.

## Section 1.3, page 16

**Exercises I**

1. a, b, f.
3. a, b, d.
5. (a) 1, 2, 3, 5, 7, 9.       (b) 2, 3, 4, 5, 6, 7, 8.
   (c) 3, 5, 7.                (d) 2.

Answers to Odd-Numbered Exercise Problems

7. $E$ and $F$.
9. (a) 5. (b) 4. (c) 6. (d) 3. (e) 4. (f) 0.

**Exercises II**

1. (a) $a, b, c, d, g, h$.    (b) $a, b$.
   (c) $a, b, c, d, e, f, g, h$.    (d) $a, b, c, d$.
3. Yes.
5. Yes.
7. (a) $e, f, g, h$.    (b) $c, d, e, f$.
   (c) $a, b, c, d, e$.    (d) $e$
9. (a) Yes.    (b) Yes.    (c) Yes.
11. Twenty-four.

## Section 1.4, page 17

1. (a) Units or ones. (b) Thousands. (c) Hundreds. (d) Millions. (e) Ten thousands.
3. Four thousand, two hundred ten.
5. Four hundred twenty-one thousand, six hundred seventy-one.
7. Seventy-one.
9. (a) Three million, five hundred forty thousand, nine hundred thirty-nine. Eight million, six hundred forty-six thousand, four hundred eighty-nine.
   (b) Five hundred sixty-six thousand, four hundred thirty-two. One thousand, forty-nine.
11. (a) Cardinal. Ordinal.    (b) Cardinal. Ordinal.
    (c) Cardinal. Ordinal.
13. (a) CMLXXXVII.    (b) MLXVII.    (c) MMCXLII.
    (d) MI.
15. a, c.
17. a, b, c, e.
19. $A$ and $B$. $A$ and $C$.

## Section 2.1, page 25

**Exercises I**

1.

| + | 0 | 1 | 2 | 3 | 4 | 5 | 6 | 7 | 8 | 9 |
|---|---|---|---|---|---|---|---|---|---|---|
| 0 | 0 | 1 | 2 | 3 | 4 | 5 | 6 | 7 | 8 | 9 |
| 1 | 1 | 2 | 3 | 4 | 5 | 6 | 7 | 8 | 9 | 10 |
| 2 | 2 | 3 | 4 | 5 | 6 | 7 | 8 | 9 | 10 | 11 |
| 3 | 3 | 4 | 5 | 6 | 7 | 8 | 9 | 10 | 11 | 12 |
| 4 | 4 | 5 | 6 | 7 | 8 | 9 | 10 | 11 | 12 | 13 |
| 5 | 5 | 6 | 7 | 8 | 9 | 10 | 11 | 12 | 13 | 14 |
| 6 | 6 | 7 | 8 | 9 | 10 | 11 | 12 | 13 | 14 | 15 |
| 7 | 7 | 8 | 9 | 10 | 11 | 12 | 13 | 14 | 15 | 16 |
| 8 | 8 | 9 | 10 | 11 | 12 | 13 | 14 | 15 | 16 | 17 |
| 9 | 9 | 10 | 11 | 12 | 13 | 14 | 15 | 16 | 17 | 18 |

3. (a) 15.  (b) 18.  (c) 18.  (d) 15.  (e) 17.
5. (a) It is the same as the headline.
   (b) It is the same as the sideline.
   (c) You start with the number in the sideline as the first entry in the first column. Thereafter, each entry increases by one.
7. (a) 6.  (b) 5.  (c) 8.  (d) No.
9. (a) Yes.  (b) Yes.

**Exercises II**

1. $((a + b) + c) + d = d + ((a + b) + c)$   Commutative property.
   $= d + (c + (a + b))$   Commutative property.
   $= d + (c + (b + a))$   Commutative property.
   $= d + ((c + b) + a)$   Associative property.
   $= (d + (c + b)) + a$   Associative property.
   $= ((d + c) + b) + a$   Associative property.

   The proof is not unique. The closure property and equality properties are used in an obvious way.

3. Let $A$ be the set who passed the first test and let $B$ be the set who passed the second test. Since no one failed both tests, $n(A \cup B) = 73$. Since $n(A) + n(B) = n(A \cup B) + n(A \cap B)$, 68

Answers to Odd-Numbered Exercise Problems      255

  $+ 65 = 73 + n(A \cap B)$; $133 = 73 + n(A \cap B)$; and $n(A \cap B) = 60$.
5. 458.
7. Yes. The sets $(P \cap Q) \cap (Q \cap R)$ and $P \cap Q \cap R$ are the same.
9. Yes because $P \cap (Q \cup R) = (P \cap Q) \cup (P \cap R)$.

## Section 2.2, page 31

### Exercises I

1. (a) 63.      (b) 132.      (c) 132.
3. (a) 1,335.   (b) 977.      (c) 699.
5. (a) 195.     (b) 240.      (c) 230.
7. (a) 13,919.  (b) 15,051.   (c) 17,141.
9. (a) 17,683.  (b) 11,410.   (c) 15,810.

### Exercises II

1. 18,668.   3. 13,053.   5. 23,842.   7. 10,127.   9. 68,713.

## Section 2.3, page 34

### Exercises I

1.  2,478  $\xrightarrow{+}$ 21 $\xrightarrow{+}$ 3
    6,889  $\xrightarrow{+}$ 31 $\xrightarrow{+}$ 4
    4,552  $\xrightarrow{+}$ 16 $\xrightarrow{+}$ 7
    ──────
    13,919 $\xrightarrow{+}$ 23 $\xrightarrow{+}$ ⑤          14 $\xrightarrow{+}$ ⑤

3.  4,665  $\xrightarrow{+}$ 21 $\xrightarrow{+}$ 3
    9,143  $\xrightarrow{+}$ 17 $\xrightarrow{+}$ 8
    3,333  $\xrightarrow{+}$ 12 $\xrightarrow{+}$ 3
    ──────
    17,141 $\xrightarrow{+}$ 14 $\xrightarrow{+}$ ⑤          14 $\xrightarrow{+}$ ⑤

5.  10,261  —+→ 10 —+→ 1
    45,872  —+→ 26 —+→ 8
    44,918  —+→ 26 —+→ 8
    191,051 —+→ 17 —+→ ⑧        17 —+→ ⑧

7.  3,891  —+→ 21 —+→ 3
    5,904  —+→ 18 —+→ 9
    3,567  —+→ 21 —+→ 3
    4,321  —+→ 10 —+→ 1
    17,683 —+→ 25 —+→ ⑦        16 —+→ ⑦

9.  8,882  —+→ 26 —+→ 8
    1,655  —+→ 17 —+→ 8
    3,258  —+→ 18 —+→ 9
    2,015  —+→ 8
    15,810 —+→ 15 —+→ ⑥        33 —+→ ⑥

11. 31,555  —+→ 19 —+→ 10 —+→ 1
    12,003  —+→ 6
    22,542  —+→ 15 —+→ 6
    13,578  —+→ 24 —+→ 6
    34,645  —+→ 22 —+→ 4
    114,323 —+→ 14 —+→ ⑤       23 —+→ ⑤

## Section 2.4, page 35

1.  7.        3. 0.         5. 17.         7. 7.         9. 4.
11. (a) 8.         (b) 13.         (c) 12.         (d) 15.
13. (a) $8 + 3 = 6 + 5$.         (b) $13 + 2 = 6 + 9$.
    (c) $12 + 2 = 5 + 9$.

Answers to Odd-Numbered Exercise Problems

15.  23,576 —+→ 23 —+→ 5
     41,784 —+→ 24 —+→ 6
     38,558 —+→ 29 —+→ 11 —+→ 2
     51,221 —+→ 11 —+→ 2
     32,569 —+→ 25 —+→ 7
     187,708 —+→ 31 —+→ ④        22 —+→ ④

17.  19,830 —+→ 21 —+→ 3
     32,477 —+→ 23 —+→ 5
     58,947 —+→ 33 —+→ 6
     11,856 —+→ 21 —+→ 3
     78,324 —+→ 24 —+→ 6
     201,434 —+→ 14 —+→ 5        23 —+→ ⑤

19.  458,977 —+→ 40 —+→ 4
     200,050 —+→ 7
     114,671 —+→ 20 —+→ 2
     456,876 —+→ 36 —+→ 9
     318,592 —+→ 28 —+→ 10 —+→ 1
     1,549,166 —+→ 32 —+→ ⑤      23 —+→ ⑤

## Section 3.1, page 39

**Exercises I**

1. (a) 8.   (b) 9.   (c) 5.   (d) 7.
3. (a) 0.   (b) The upper right-hand part.
5. 4.    7. 10.    9. 0.

**Exercises II**

1. (a) 6.   (b) 10.
3. (a) $A \cup B = B$.   (b) $A \cap B = A$.

5. (a) Yes.               (b) Yes.
7. $n(S' \cup S) + n(S' \cap S) = n(S') + n(S)$
   $n(T) + n(\emptyset) = n(S') + n(S)$
   $n(T) + 0 = n(S') + n(S)$
   $n(T) - n(S) = n(S')$.
9. Yes.

## Section 3.2, page 42

**Exercises I**

1. 246.          3. 417.          5. 345.
7. 2,060.        9. 14,825.       11. 308,977.

**Exercises II**

1. 2,440.        3. 45,251.       5. 48,549.
7. 125,419.      9. 1,990,888.

11. $43{,}566{,}870 \xrightarrow{+} 39 \xrightarrow{+} 12 \xrightarrow{+} \boxed{3}$
    $21{,}823{,}445 \xrightarrow{+} 29 \xrightarrow{+} 11 \xrightarrow{+} 2$
    $21{,}743{,}425 \xrightarrow{+} 28 \xrightarrow{+} 10 \xrightarrow{+} 1$
    $\phantom{21{,}743{,}425 \xrightarrow{+} 28 \xrightarrow{+} 10 \xrightarrow{+} \;} \boxed{3}$

## Section 3.3, page 44

**Exercises I**

1. a, b, d, e.
3. 3 is the least element and 23 is the greatest.
5. 0 is the least element and 7 is the greatest.
7. 15 is the least element. The set has no greatest element.
9. 3 is the least element and 32 is the greatest.

**Exercises II**

1. 1 is the least element and 16 is the greatest.
3. 5 is the least element and 31 is the greatest.

Answers to Odd-Numbered Exercise Problems

5. Yes. This follows from the definition of greater-than and the trichotomy property.
7. If $a < b$, then there exists a natural number $x$ such that $a + x = b$. By the closure and equality properties, $(a + x) + c = b + c$; thus $(a + c) + x = b + c$ by the associative and commutative properties. Since $x$ is a natural number, $a + c < b + c$ by the definition of less-than.
9. If $a < b$, we can conclude from Exercise 7 that $a + c < b + c$. If $c < d$, we can conclude from Exercise 7 that $c + b < d + b$, and from the commutative property that $b + c < b + d$. From $a + c < b + c$ and $b + c < b + d$, we conclude that $a + c < b + d$ from the transitive property.

## Section 3.4, page 45

1. a, b, c, e, f.
3. 1 is the least element and 14 is the greatest element.
5. 36 is the least element. The set has no greatest element.
7. Yes.       9. 35.         11. 38.          13. 24,245.
15. 8,860.    17. 530,865.   19. 646,653.

## Section 4.1, page 53

**Exercises I**

1. (a) 24.    (b) 56.    (c) 54.    (d) 0.    (e) 14.
3. (a) 0.     (b) The entries are the same.    (c) Yes.
5. (a) Yes, it is 3.    (b) No.    (c) No.
7. (a) No.    (b) Yes.    (c) 20, 40, 60.
9. (a) 6, 12, 18, 24, 30, 36, 42, 48, 54, 60.
   (b) 15, 30, 45, 60, 75, 90, 105, 120, 135, 150.
   (c) 30, 60, 90.
   (d) 30.

**Exercises II**

1. (a) Because $a \times b$ is both a multiple of $a$ and a multiple of $b$.
   (b) The well-ordering property.

3. Since $u$ is a multiple of $a$ there exists a natural number $x$ such that $u = ax$. Similarly, there is a natural number $y$ such that $v = ay$. Now, $u + v = ax + ay = a(x + y)$ by the distributive property. From the closure property for addition, $x + y$ is a natural number. By definition, $u + v$ is a multiple of $a$.

5. (a) 6, 12, 18, 24, 30, 36, 42.   (b) 42.   (c) 5.   (d) 7.

7. No. $2 < 5$ but $2 \times 0$ is not less than $5 \times 0$.

9. If $a$ is a multiple of $c$ then there is a natural number $x$ such that $a = cx$. Hence, $ab = (cx)b = c(xb)$. Since $x$ and $b$ are natural numbers, $xb$ is a natural number, and $ab$ is a multiple of $c$ by the definition of multiple.

## Section 4.2, page 59

**Exercises I**

1. 9,468.  3. 7,872.  5. 28,770.
7. 67,362.  9. 495,000.  11. 29,760,408.

**Exercises II**

1. 247,364.  3. 15,272.  5. 471,000.
7. 3,118,356.  9. 256,844,126.  11. 3,756,913,104.

## Section 4.3, page 61

**Exercises I**

1. 1, 2, 4, 8, 16, 32.
3. 1, 23.
5. 1, 2, 4, 5, 10, 20, 25, 50, 100.
7. (a) 56.   (b) 6.
9. (a) 90.   (b) 8.

**Exercises II**

1. (a) 6.   (b) 12.   (c) 72.
3. (a) 7.   (b) 24.   (c) 168.
5. (a) 8.   (b) 12.   (c) 96.

*Answers to Odd-Numbered Exercise Problems*

7. (a) 15.  (b) 18.  (c) 186.
9. (a) $a|b$ implies there is a natural number $x$ such that $ax = b$. $a|c$ implies there is a natural number $y$ such that $ay = c$. Thus, $(ax)(ay) = bc$ and $a(axy) = bc$. Since $axy$ is a natural number, $a|bc$.
   (b) In general, no. For example, $4|(6 \times 10)$ but $4 \!\not|\, 6$ and $4 \!\not|\, 10$.

## Section 4.4, page 62

1. 6.   3. 8.   5. 6.   7. 4.
9. (a) $\{1, 2, 3, 4, 6, 8, 12, 24\}$.
   (b) $\{1, 2, 4, 5, 10, 20\}$.
11. (a) $\{1, 2, 4\}$.   (b) 1.   (c) 4.
13. (a) 7, 14, 21, 28, 35, 42, 49, 56, 63, 70, 77, 84.
    (b) 84.   (c) 0.   (d) 12.
    (e) For a set with 84 elements, the number of subsets which contain exactly 7 elements is 12 with no elements left over.
15. (a) 4.   (b) 12.   (c) 48.   (d) 2.   (e) 2.   (f) 4.
17. sum: 36,314;   difference: 11,378;   product: 297,311,928.
19. sum: 833,291;   difference: 32,641;   product: 173,327,113,950.

## Section 5.1, page 69

**Exercises I**

1. (a) 9.   (b) 4.   (c) 0.   (d) Does not exist.
   (e) No quotient in the set of whole numbers.   (f) 9.
   (g) Does not exist.   (h) 8.
3. 2, 4, 5, 8, 10, 40.
5. None of them.
7. 2, 3, 4, 5, 6, 8, 10, 11, 12, 15, 40.
9. 2, 3, 4, 6, 8, 9, 12, 18.

**Exercises II**

1. (1) Definition of positional notation.
   (2) Addition facts.
   (3) Distributive and associative properties of addition.

(4) Commutative and associative properties of addition.
(5) Distributive property.
(6) Since both $9(111a + 11b + c)$ and $(a + b + c + d)$ have 9 as a factor, so does their sum.
3. (a) 1, 2, 4, 5, 10, 20.
   (b) 1, 3, 5, 9, 15, 45.
   (c) 20, 40, 60, 80, 100, 120, 140, 160, 180, 200.
   (d) 45, 90, 135, 180, 225, 270, 315, 360, 405, 450.
5. $G \times L = 5 \times 180 = 900$.
7. (a) 1, 2, 3, 6, 9, 18.
   (b) 1, 2, 3, 5, 6, 10, 15, 30.
   (c) 18, 36, 54, 72, 90, 108, 126, 144, 162, 180.
   (d) 30, 60, 90, 120, 150, 180, 210, 240, 270, 300.
9. $G \times L = 6 \times 90 = 540$.

## Section 5.2, page 74

**Exercises I**

1. $q = 140, r = 32$.
3. $q = 48, r = 46$.
5. $q = 144, r = 159$.
7. $q = 1328, r = 395$.
9. No.

**Exercises II**

1. Let $q = 0$ and $r = a$.
3. (b) $x = 1,615$.
5. $q = 244, r = 317$.
7. $q = 994, r = 93$.
9. $q = 1,698, r = 666$.
11. Since $a \div b = x$ and $c \div d = x$, it follows that $bx = a$ and $dx = c$. Thus, $d(bx) = ad$ and $b(dx) = bc$, and we conclude that $ad = bc$.
13. From $a \div (b \div c) = 1$ we conclude that $a = (b \div c) \times 1$; thus, $a = b \div c$. Consequently, $ac = b$.

## Section 5.3, page 79

**Exercises I**

1. 2,050.
3. 1,056.
5. 1,512.
7. 2,142.
9. 817.

*Answers to Odd-Numbered Exercise Problems* 263

**Exercises II**

1. 3,108.   3. 2,166.   5. 11,928.
7. 32,562.   9. 31,512.

## Section 5.4, page 80

1. (a) Addition, subtraction, multiplication, and division.
   (b) See text.
3. 100.   5. 55.   7. 82.
9. 223.   11. (b) $x = 2,131$.   13. (b) $x = 1,892,467$.
15. 7,046.   17. 6,111.   19. 3,071.

## Section 6.1, page 86

**Exercises I**

1. (a) 2, 3, 5, 7, 11, 13, 17, 19, 23, 29, 31, 37, 41, 43, 47, 53, 59, 61, 67, 71, 73, 79, 83, 89, 97, 101, 103, 107, 109, 113, 127, 131, 137, 139, 149, 151, 157, 163, 167, 173, 179, 181, 191, 193, 197, 199.
   (b) 46.
3. 3 and 5, 5 and 7, 11 and 13, 17 and 19, 29 and 31, 41 and 43, 59 and 61, 71 and 73, 101 and 103, 107 and 109, 137 and 139, 149 and 151, 179 and 181, 191 and 193, 197 and 199. (It is unknown whether the number of such pairs is finite or infinite.)
5. (a) No.   (b) Yes.
7. (a) Yes.   (b) No.
9. $N = 1$; $(N \times N) + 1 = 2$ is a prime.
   $N = 2$; $(N \times N) + 1 = 5$ is a prime.
   $N = 4$; $(N \times N) + 1 = 17$ is a prime.
   $N = 6$; $(N \times N) + 1 = 37$ is a prime.
   $N = 10$; $(N \times N) + 1 = 101$ is a prime.
   Notice for $N \neq 1$, $N$ must be an even natural number.

**Exercises II**

1. Yes.
3. (a) 2.   (b) 2.   (c) 4.
5. $d(mn) = d(m) \times d(n)$ if $m$ and $n$ are relatively prime.

7. $\phi(5) = 4$.
9. $\phi(15) = 8$.
11. $\phi(28) = 12$.
13. $\phi(40) = 16$.
15. $\phi(60) = 16$.
17. Yes.

## Section 6.2, page 88

### Exercises I

1. $360 = 2 \times 2 \times 2 \times 3 \times 3 \times 5$.
3. $366 = 2 \times 7 \times 19$.
5. $1{,}240 = 2 \times 2 \times 2 \times 5 \times 31$.
7. $6{,}946 = 2 \times 23 \times 151$.
9. $9{,}360 = 2 \times 2 \times 2 \times 2 \times 3 \times 3 \times 5 \times 13$.
11. $720 = 2 \times 2 \times 2 \times 2 \times 3 \times 3 \times 5$.

### Exercises II

1. No. 1 could be expressed as the product of primes in more than one way; for example, $1 \times 1 = 1 \times 1 \times 1$.
3. $475 = 5 \times 5 \times 19$.
5. $1{,}275 = 3 \times 5 \times 5 \times 17$.
7. $23{,}248 = 2 \times 2 \times 2 \times 2 \times 1{,}453$.
9. $6{,}080 = 2 \times 2 \times 2 \times 2 \times 2 \times 2 \times 5 \times 19$.

## Section 6.3, page 93

### Exercises I

1. 6 and 1,326.
3. 1 and 851.
5. 11 and 33.
7. 30 and 18,630.
9. 33 and 2,356,200.

### Exercises II

1. 35 and 420.
3. 1 and 4,828.
5. 1 and 1,763.
7. 1 and 13,680.
9. 29 and 1,160.
11. 31 and 28,210.

## Section 6.4, page 98

### Exercises I

1. 56.
3. 29.
5. 67.

*Answers to Odd-Numbered Exercise Problems* 265

7. 68.
9. 111.
11. 143,640.
13. 52,080.
15. 200,970.
17. 337,881.

## Section 6.5, page 98

1. 3 and 7, 7 and 11, 19 and 23, 37 and 41, 43 and 47, 67 and 71, 79 and 83, 97 and 101, 103 and 107, 109 and 113, 127 and 131, 163 and 167, 193 and 197.
3. (a) Yes.   (b) No.
5. (a) Yes.   (b) No.
7. $2 \times 2 \times 3 \times 5 \times 5$.
9. $2 \times 13 \times 41$.
11. $2 \times 5 \times 1{,}633$.
13. 6.    15. 63.    17. 38.
19. (13) – 468, (14) – 22,248, (15) – 27,342, (16) – 3,713, (17) – 9,690, (18) – 62,744.

## Section 7.1, page 105

### Exercises I

1. (a) $\frac{4}{5}$.   (b) $\frac{7}{5}$.   (c) $\frac{2}{5}$.   (d) $\frac{2}{4}$.
3. (a) $\frac{7}{10}$.   (b) $\frac{6}{10}$.   (c) $\frac{4}{10}$.   (d) $\frac{14}{10}$.
5. (a) $\frac{2}{4}$.   (b) $\frac{3}{4}$.
7. (a) $\frac{5}{16}$.   (b) $\frac{7}{16}$.
9. (a) $\frac{16}{10}, \frac{8}{5}$.   (b) $\frac{2}{4}, \frac{1}{2}$.   (c) $\frac{6}{4}, \frac{3}{2}$.   (d) $\frac{1}{4}, \frac{2}{8}$.
   (e) $\frac{12}{4}, \frac{6}{2}$.   (f) $\frac{6}{4}, \frac{3}{2}$.   (g) $\frac{5}{2}, \frac{10}{4}$.   (h) $\frac{2}{6}, \frac{1}{3}$.
   (i) $\frac{5}{3}, \frac{10}{6}$.   (j) $\frac{4}{5}, \frac{8}{10}$.   (k) $\frac{7}{5}, \frac{14}{10}$.

### Exercises II

1. $\frac{6}{4}$.   3. $\frac{3}{6}$.   5. $\frac{7}{5}$.   7. $\frac{5}{3}$.   9. $\frac{3}{7}$.

## Section 7.2, page 111

### Exercises I

1. (a) $\frac{11}{15}$.   (b) Equal.
3. (a) Equal.   (b) $\frac{63}{71}$.

5. (a) $\frac{49}{91}$. (b) Equal.
7. (a) $\{\frac{49}{70}, \frac{18}{70}\}$. (b) $\{\frac{44}{143}, \frac{65}{143}\}$.
9. (a) $\{\frac{989}{1247}, \frac{174}{1247}\}$. (b) $\{\frac{42}{68}, \frac{41}{68}\}$.
11. (a) $\frac{5}{11}$. (b) $\frac{3}{2}$. (c) $\frac{4}{7}$. (d) $\frac{7}{11}$.

**Exercises II**

1. (a) $\frac{2}{3}$. (b) $\frac{91}{36}$.
3. (a) $\frac{17}{12}$. (b) $\frac{129}{260}$.
5. (a) $\frac{3}{7}$. (b) Equal.
7. (a) $\{\frac{10}{45}, \frac{42}{45}\}$. (b) $\{\frac{51}{187}, \frac{22}{187}\}$.
9. (a) $\{\frac{25}{70}, \frac{24}{70}\}$. (b) $\{\frac{392}{1,127}, \frac{138}{1,127}\}$.
11. "Since the numerator of $\frac{6}{8}$" refers to the symbol (fraction) "$\frac{6}{8}$". It is not true that the symbol "$\frac{6}{8}$" is equal to the symbol "$\frac{3}{4}$"; that is, "$\frac{6}{8}$" $\neq$ "$\frac{3}{4}$". Of course, $\frac{6}{8} = \frac{3}{4}$, but this equality is for numbers and not symbols.

## Section 7.3, page 112

1. (a) $\frac{11}{19}$. (b) $\frac{16}{19}$.
3. (a) $\frac{16}{5}$. (b) $\frac{12}{5}$.
5. (a) $\frac{2}{5}$. (b) $\frac{5}{6}$.
7. (a) $\{\frac{27}{45}, \frac{20}{45}\}$. (b) $\{\frac{20}{40}, \frac{16}{40}, \frac{15}{40}\}$.
9. (a) $\{\frac{15}{24}, \frac{14}{24}\}$. (b) $\{\frac{36}{234}, \frac{117}{234}, \frac{104}{234}\}$.
11. (a) Equal. (b) $\frac{13}{9}$.
13. (a) $\frac{16}{27}$. (b) $\frac{53}{71}$.
15. (a) $\frac{481}{322}$. (b) Equal.

## Section 8.1, page 119

**Exercises I**

1. (a) $\frac{37}{55}$. (b) $\frac{67}{56}$.
3. (a) $\frac{15}{22}$. (b) $\frac{11}{12}$.
5. (a) $\frac{89}{90}$. (b) $\frac{31}{30}$.
7. (a) $\frac{11}{45}$. (b) $\frac{95}{84}$.
9. (a) $\frac{69}{40}$. (b) $\frac{37}{12}$.

*Answers to Odd-Numbered Exercise Problems*

**Exercises II**

1. $\dfrac{a}{b} + 0 = \dfrac{a}{b} + \dfrac{0}{d} = \dfrac{ad + b \cdot (0)}{bd} = \dfrac{ad + 0}{bd} = \dfrac{ad}{bd} = \dfrac{a}{b}.$

3. $\dfrac{104}{63}.$   5. $\dfrac{31}{22}.$   7. $\dfrac{35}{12}.$   9. $\dfrac{3}{2}.$

## Section 8.2, page 123

**Exercises I**

1. (a) $\dfrac{26}{7}.$   (b) $\dfrac{44}{9}.$   (c) $\dfrac{71}{11}.$   (d) $\dfrac{61}{8}.$
3. (a) $11\tfrac{1}{6}.$   (b) $15\tfrac{7}{10}.$   (c) $13\tfrac{8}{21}.$
5. (a) $14\tfrac{7}{16}.$   (b) $28\tfrac{61}{66}.$   (c) $59\tfrac{17}{60}.$
7. (a) $6\tfrac{1}{6}.$   (b) $3\tfrac{11}{42}.$   (c) $6\tfrac{1}{2}.$
9. (a) $3\tfrac{13}{45}.$   (b) $2\tfrac{13}{24}.$   (c) $6\tfrac{13}{45}.$

**Exercises II**

1. (a) $\dfrac{21}{5}.$   (b) $\dfrac{26}{9}.$   (c) $\dfrac{87}{15}.$   (d) $\dfrac{54}{13}.$
3. (a) $13\tfrac{10}{17}.$   (b) $8\tfrac{2}{5}.$   (c) $4\tfrac{13}{17}.$   (d) $3\tfrac{17}{23}.$
5. $21\tfrac{7}{10}.$   7. $34\tfrac{19}{42}.$   9. $12\tfrac{151}{168}.$

## Section 8.3, page 128

**Exercises I**

1. (a) $2\tfrac{1}{2}.$   (b) $4\tfrac{1}{4}.$   (c) $2\tfrac{1}{8}.$
3. (a) $5\tfrac{5}{8}.$   (b) $7\tfrac{3}{5}.$   (c) $10\tfrac{3}{25}.$
5. (a) 56.21.   (b) 107.53.   (c) 8.199.
7. (a) 11.347.   (b) 14.827.   (c) 21.080.
9. (a) 45.102.   (b) 36.355.   (c) 4.9538.

**Exercises II**

1. 41.31.   3. 16.5166.   5. 32.951.
7. 26.402.   9. 7.008.

## Section 8.4, page 134

**Exercises I**

1. a, c, and d.
3. (a) 0.62.　　(b) 0.54.　　(c) 0.73.　　(d) 0.13.
5. (a) 0.625.　　(b) 0.273.　　(c) 0.444.　　(d) 0.917.
7. (a) 0.8750.　　(b) 0.4167.　　(c) 0.0236.　　(d) 0.0057.
9. (a) 0.18248.　　(b) 0.01646.　　(c) 0.20190.　　(d) 0.00642.

**Exercises II**

1. (a) 0.5.　　(b) 0.6.　　(c) 3.6.　　(d) 6.5.
3. (a) 0.294.　　(b) 0.320.　　(c) 0.687.　　(d) 7.077.
5. (a) 5.85714.　　(b) 0.19315.　　(c) 0.15000.　　(d) 0.78846.
7. (a) 0.6667.　　(b) 0.5000.　　(c) 0.8476.　　(d) 0.7825.
9. 2.7967.

## Section 8.5, page 135

1. $1\frac{27}{40}$.　　3. $20\frac{13}{24}$.　　5. $18\frac{2}{5}$.　　7. 0.6296.　　9. 0.9007.

## Section 9.1, page 142

**Exercises I**

1. (a) $\frac{10}{21}$.　　(b) $\frac{15}{88}$.　　(c) $\frac{6}{7}$.　　(d) $\frac{2}{5}$.
3. (a) $1\frac{1}{8}$.　　(b) $1\frac{1}{5}$.　　(c) $\frac{21}{34}$.　　(d) $1\frac{7}{11}$.
5. (a) $\frac{10}{121}$.　　(b) $1\frac{137}{183}$.　　(c) $\frac{3}{10}$.　　(d) 6.
7. (a) $\frac{5}{16}$.　　(b) $1\frac{43}{55}$.　　(c) $\frac{4}{121}$.　　(d) $\frac{28}{45}$.
9. (a) $\frac{6}{7}$.　　(b) $\frac{4}{63}$.　　(c) $\frac{17}{55}$.　　(d) $1\frac{91}{121}$.

**Exercises II**

1. Step 1. Definition of fractional notation.
　Step 2. Theorem on division of rationals written as fractions.
　Step 3. Definition of multiplication of rationals written as fractions.
　Step 4. Multiplication properties for whole numbers.
3. $3\frac{19}{54}$.　　5. $\frac{135}{224}$.　　7. $\frac{74}{135}$.　　9. $1\frac{103}{132}$.

*Answers to Odd-Numbered Exercise Problems*　　　　　　　　　　　　　　269

## Section 9.2, page 146

**Exercises I**

1. 492.703.    3. 4.390191.    5. 0.040885.
7. 1.904.     9. 3.110.

**Exercises II**

1. 1,678.6119    3. 0.1508311
5. 0.027196     7. 17.19130
9. 0.00155

## Section 9.3, page 149

**Exercises I**

1. (a) $0.36\frac{1}{6}$.   (b) $0.15\frac{19}{45}$.   (c) $0.091\frac{5}{7}$.
3. (a) $0.05\frac{1}{6}$.   (b) $0.09\frac{13}{15}$.   (c) $0.076\frac{2}{7}$.
5. (a) 37.44.      (b) 90.1.       (c) 8.9.
7. (a) 11,000.     (b) 140.        (c) 36,000.
9. (a) $\frac{31}{157}$.   (b) $\frac{13}{82}$.

**Exercises II**

1. $4.01\frac{1}{6}$.   3. $2.88\frac{7}{40}$.   5. 18.69.   7. $\frac{248}{585}$.   9. $1\frac{3}{28}$.

## Section 9.4, page 151

1. 17.3163.    3. 50.406620.    5. 325.6902.
7. 13.448.    9. 3.143.

## Section 10.1, page 157

**Exercises I**

1. (a) $-5$.    (b) $-53$.    (c) 14.
3. (a) $-2$.   (b) 16.       (c) 0.

5. (a) $-\frac{7}{16}$. (b) $\frac{7}{40}$. (c) $-\frac{43}{72}$.
7. (a) $-41$. (b) 35. (c) $-46$.
9. (a) $-\frac{59}{42}$. (b) $\frac{13}{72}$. (c) $\frac{53}{48}$.
11. (a) $-1.804$. (b) 1.43. (c) 0.7.

**Exercises II**

1. $a + (b + c) = (-3) + [(-7) + 8] = (-3) + 1 = -2$.
   $(a + b) + c = [(-3) + (-7)] + 8 = (-10) + 8 = -2$.
3. $a + (b + c) = (-8) + [4 + (-11)] = (-8) + (-7) = -15$.
   $(a + b) + c = [(-8) + 4] + (-11) = (-4) + (-11) = -15$.
5. $-111$.      7. $-11,557$.      9. (a) $-\frac{101}{63}$.    (b) $\frac{71}{60}$.


## Section 10.2, page 160

**Exercises I**

1. (a) $-54$.      (b) $-115$.
3. (a) $-4$.       (b) 3.
5. (a) $-\frac{3}{8}$.  (b) $-\frac{23}{22}$.
7. (a) $-\frac{8}{15}$. (b) $\frac{16}{35}$.
9. (a) $-\frac{1}{20}$. (b) $\frac{11}{90}$.

**Exercises II**

1. $a(b + c) = (-3)[(-7) + 8] = (-3)(1) = -3$.
   $ab + ac = (-3)(-7) + (-3)(8) = 21 + (-24) = -3$.
3. $a(b + c) = (-8)[4 + (-11)] = (-8)(-7) = 56$.
   $ab + ac = (-8)(4) + (-8)(-11) = (-32) + 88 = 56$.
5. $\frac{11}{6}$.     7. 36.33.      9. $-0.433$.


## Section 10.3, page 161

**Exercises I**

1. True.     3. False.     5. True.     7. False.     9. False.

**Exercises II**

1. No.     3. Yes.
5. If $a < b$, then by Exercise 5, $a + c < b + c$. Similarly, if $c < d$,

then $c + b < d + b$ by adding $b$ to both sides of the inequality. Thus, by the commutative law, $b + c < b + d$. Since $a + c < b + c$ and $b + c < b + d$, by the transitive property we have $a + c < b + d$.

7. Let us give an indirect proof. By the trichotomy property, if $\frac{a}{b} < \frac{c}{d}$ is false then $\frac{a}{b} \geq \frac{c}{d}$. If $\frac{a}{b} \geq \frac{c}{d}$, then multiplying both sides by the positive number $bd$ we get $ad \geq bc$. This contradicts the assumption that $ad < bc$.

9. Let us give an indirect proof. If $\frac{1}{a} > \frac{1}{b}$ is not true then $\frac{1}{a} \leq \frac{1}{b}$. Multiplying both sides by the positive number $ab$ we obtain that $b \leq a$, but this contradicts the assumption that $a < b$.

## Section 10.4, page 162

1. (a) $-16$.    (b) $14$.    (c) $-85$.
3. (a) $-126$.   (b) $-20$.   (c) $21$.
5. $-2.69$.      7. $11.985$. 9. $-10.017$.

## Section 11.2, page 169

**Exercises I**

1. (a) 100 meters.        (b) 337 inches.
3. (a) 10,000,000 meters. (b) 40,000,000 meters.
   (c) 40,000,000 meters is equivalent to 40,000 kilometers.
       From example 3, 1 kilometer = $3,280\frac{5}{6}$ ft. Thus,

       $$40,000 \text{ kilometers} = 40,000 \times 3,280\tfrac{5}{6} \text{ ft}$$

       $$= 40,000 \times \frac{19,685}{6}$$

       $$= 20,000 \times \frac{19,685}{3} \text{ ft.}$$

   Since 5,280 ft = 1 mile,

   $$40,000 \text{ kilometers} = \frac{20,000 \times 19,685}{5,280 \times 3} \text{ miles}$$

   $$= 24,817 \text{ miles to the nearest mile.}$$

5. (a) 15 meters × 100 $\frac{\text{cm}}{\text{meter}}$ = 1,500 cm.

   (b) 15 feet × 12 $\frac{\text{in.}}{\text{ft}}$ = 180 in.

7. (a) 537 cm = 5.37 meters.
   (b) 537 inches = $\frac{537}{12}$ ft = $44\frac{3}{4}$ ft, or 44.75 ft.
9. (a) 5,280 ÷ $16\frac{1}{2}$ = 5,280 × $\frac{2}{33}$ = $\frac{10,560}{33}$ = 320 rods.
   (b) 38 yards = 38 × 3 feet = 114 ÷ $16\frac{1}{2}$ rods. Thus, 38 yards = 114 × $\frac{2}{33}$ = $\frac{228}{33}$ = $6\frac{10}{11}$ rods.

## Section 11.3, page 175

### Exercises I

1. $\frac{1}{640}$ sq mi = 1 acre and 5,280 × 5,280 sq ft = 1 sq mi. Thus 1 acre = $\frac{1}{640}$ × 5,280 × 5,280 sq ft = 43,560 sq ft.
3. (See Exercise 1.) $16\frac{1}{2}$ × $16\frac{1}{2}$ sq ft = $\frac{1,089}{4}$ sq ft = 1 sq rod. Thus, 1 acre = 43,560 ÷ $\frac{1,089}{4}$ sq rods = 160 sq rods.
5. Area = $25\frac{1}{2}$ × $11\frac{1}{4}$ = $\frac{51}{2}$ × $\frac{51}{4}$ = $286\frac{7}{8}$ sq in.
7. 1 ft 2 in. = 14 in. Area = $\frac{1}{2}$ × 14 × $9\frac{5}{6}$ = $68\frac{5}{6}$ sq in.
9. 2 ft 5 in. = 29 in. Area = $\frac{1}{2}$(29 + $42\frac{2}{3}$)38 = $1,361\frac{2}{3}$ sq in.
11. (a) 20 ft 6 in. = (20 × 12) + 6 = 246 in. and 12 ft 3 in. = (12 × 12) + 3 = 147 in. Thus, area = 147 × 246 = 36,162 sq in.
    (b) 36,162 × $\frac{1}{144}$ = 251.125 sq ft.
    (c) 251.125 × $\frac{1}{9}$ = 27.90$\overline{27}$ ... sq yd.
13. Area = length × width and $4\frac{2}{3}$ yd = 14 ft. Thus, 124 = 14 × width and width = 124 ÷ 14 = $8\frac{6}{7}$ ft.
15. (a) 1 ft 2 in. = 14 in. Thus, area = $\pi r^2$ = $(14)^2 \pi$ = 615.44 sq in. where $\pi$ = 3.14.
    (b) 2 ft 2 in. = 26 in. Circumference = $2\pi r$ = $52\pi$ = 163.28 in. where $\pi$ = 3.14.
17. Perimeter = 17 ft 6 in.
19. 8 × (1 ft 8 in.) = 8 ft 64 in. = 13 ft 4 in.
21. $2w + 2l$ = perimeter; 2 ft 7 in. = 31 in. Thus, 2(31) + $2l$ = 412, $2l$ = 350, $l$ = 175 in.

Answers to Odd-Numbered Exercise Problems

## Section 11.4, page 179

### Exercises I

1. Volume $= 3\frac{1}{2} \times 2\frac{1}{3} \times 6 = \frac{7}{2} \times \frac{7}{3} \times 6 = 49$ cu ft.
   Surface area $= 2(6 \times 2\frac{1}{3}) + 2(3\frac{1}{2} \times 2\frac{1}{3}) + 2(3\frac{1}{2} \times 6)$
   $= 86\frac{1}{3}$ sq ft.
3. 2 ft 5 in. $= 29$ in. and $3\frac{1}{6}$ ft. $= 38$ in. Volume $= 12,122$ cu in.; surface area $= 3,678$ sq in.
5. 2 ft 3 in. $= 27$ in. and 3 ft 5 in. $= 41$ in. Volume $= 29,889\pi = 93,851.46$ cu in. Surface area $= 3,672\pi = 11,530.08$ sq in.
7. Volume $= 66.98\frac{2}{3} = 66.986\ldots$ cu in.
9. Volume $= 243.42$ cu in.
11. Surface area $= 567$ sq ft.
13. $(2.54)^3$ cu cm $= 1$ cu in.; thus, 1 cu in. $= 16.387064$ cu cm.
15. $14,649,984$ sq mi.
17. Eight.
19. $360\pi = 1,130.4$ cu in. (approx.).

## Section 11.5, page 185

### Exercises I

1. (a) 14 qt.  (b) 28 qt.
3. (a) 2,400 minutes.  (b) 144,000 seconds.
5. 12 miles per hour.
7. 1,188 degrees.
9. $t_1 = 40 \div 75 = \frac{8}{15}$ hr, $t_2 = 40 \div 70 = \frac{4}{7}$ hr. $t_2 - t_1 = \frac{4}{105}$ hr $= 2.286$ min (approx.).

## Section 11.6, page 188

### Exercises I

1. $38 \div 3\frac{3}{8} = 11\frac{7}{27}$ cents per ounce. $79 \div 7 = 11\frac{2}{7}$ cents per ounce. Since $\frac{7}{27} < \frac{2}{7}$, the smaller jar is the better buy.
3. 11 ft 6 in. $= 138$ in. and 18 ft $= 216$ in. Thus, the floor area is

29,808 sq in. Since each tile is 81 sq in., it will require 29,808 ÷ 81 = 368 tiles. The cost is 368 × $19\frac{1}{2}$ = 7,176 cents, or $71.76.
5. 300 ÷ $15\frac{1}{2}$ = $19.3\frac{17}{31}$. Thus, 19.4 gallons (to the nearest tenth) are needed. The price is 38.9 × 19.4 = 754.66 cents. The cost will be $7.55.
7. Front yard area = 2400 sq ft and back yard = 5225 sq ft. Total area = 7,625 sq ft. The amount needed is

$$\frac{40 \text{ lb}}{10,000 \text{ sq ft}} \times 7,625 \text{ sq ft} = 30.5 \text{ lb.}$$

Total cost is 30 × 30.5 = 915 cents = $9.15.
9. Let $x$ be the cost of Type A. Then $486 - x$ is cost of Type B. Hence,

$$5x + 2(486 - x) < 1800$$
$$5x + 972 - 2x < 1800$$
$$3x < 1800 - 972$$
$$3x < 828$$
$$x < 276.$$

Maximum cost is $275.99.

## Section 11.7, page 195

**Exercises I**

1. (a) 50th term: $4 + (49)5 = 249$.
   (b) $S = \frac{50}{2}[8 + (49)5] = 6,325$.
3. (a) 50th term: $3 + (49)\frac{4}{3} = \frac{205}{3} = 68\frac{1}{3}$.
   (b) $S = \frac{50}{2}[6 + (49)\frac{4}{3}] = \frac{5,350}{3} = 1,783\frac{1}{3}$.
5. (a) $d = 3$; $S = \frac{30}{2}[16 + (29)3] = 1,545$; 50th term = 155.
   (b) $y = 8\frac{1}{2}$; $d = 3\frac{1}{2}$; $S = \frac{20}{2}[5 + (19)\frac{7}{2}] = 715$.
7. (a) $18 = 8 + 4d$, $d = \frac{5}{2}$. Thus, $s = \frac{21}{2}$, $t = 13$, and $u = \frac{31}{2}$.
   $S = \frac{10}{2}[16 + (9)\frac{5}{2}] = \frac{385}{2} = 192\frac{1}{2}$.
   (b) $d = 4$; $97 = -3 + (n - 1)4$; $4n = 104$, $n = 26$. Thus, $S = \frac{26}{2}[-3 + 97] = 1,222$.
9. $8300 + 15(500) = 15,800$ dollars.

*Answers to Odd-Numbered Exercise Problems*

## Section 11.8, page 201

1. $r = -1$, $a_7 = 3$.
3. $r = 3$, $a_7 = 1{,}458$.
5. $r = -\frac{1}{3}$, $a_7 = \frac{1}{9}$.
7. (a) $A = 1{,}000[1 + \frac{0.05}{4}]^4 = 1{,}000(1 + 0.0125)^4$
   $= 1{,}000(1.0125)^4$.
   Using the table on page 285, $A \approx 1{,}000(1.05095) = \$1{,}050.95$.
   (b) 5.095 percent.
9. $2P = P[1 + \frac{0.05}{4}]^{4n}$; thus, $(1.0125)^{4n} = 2$. Using the table on page 285, $(1.0125)^{56} = 2.005^+$. Thus, $4n = 56$ and $n = 14$ years.
11. $P(1 + 0.05)^n = P[1 + \frac{0.03}{4}]^{4(10)}$; thus, $(1.05)^n = (1.0075)^{40}$.
    Using the table on page 283, $(1.0075)^{40} = 1.34835$. Also, $(1.05)^n \approx 1.34835$ for $n = 6^+$.

## Section 12.1, page 206

### Exercises I

1. 2, 3, 4, 5, 6, 8, 9, 10, 11, 15, 18, and 40.
3. 2, 3, 4, 5, 6, 8, 9, 10, 15, 18, and 40.
5. 2, 4, 5, and 10.
7. 3, 9, and 11.
9. 2, 3, 4, 5, 6, 8, 10, 11, 15, and 40.

### Exercises II

1. From Part 1,
   $(a + b + c + d + e + f) = abc{,}def - 3[33{,}333a + 3{,}333b + 333c + 33d + 3e]$. If 3 is a factor of $abc{,}def$, then 3 is a factor of the difference on the right-hand side of the equality; thus, 3 is a factor of $(a + b + c + d + e + f)$.
3. Hint: $abc{,}def = 10^5 a + 10^4 b + 10^4 c + 10e + f$
   $= 5[2(10^4)a + 2(10^3)b + 2(10^2)c + 2(10)d + 2e] + f$.
5. Use $abc{,}def = [1000(10^2)a + 1000(10)b + 1000c]$
   $+ [10^2 d + 10e + f]$
   $= 8[125(10^2)a + 125(10)b + 125c]$
   $+ [10^2 + 10e + f]$.
7. Hint: $abc{,}def = 10[(10^4)a + (10^3)b + (10^2)c + 10d + e] + f$.

## Section 12.2, page 208

### Exercises I

1.  $4 = 2 + 2$,  $6 = 3 + 3$,  $8 = 3 + 5$,  $10 = 3 + 7$,
    $12 = 5 + 7$,  $14 = 3 + 11$,  $16 = 3 + 13$,  $18 = 5 + 13$,
    $20 = 3 + 17$,  $22 = 3 + 19$,  $24 = 5 + 19$,  $26 = 3 + 23$,
    $28 = 5 + 23$,  $30 = 7 + 23$,  $32 = 3 + 29$,  $34 = 5 + 29$,
    $36 = 5 + 31$,  $38 = 7 + 31$,  $40 = 3 + 37$,  $42 = 5 + 37$,
    $44 = 7 + 37$,  $46 = 3 + 43$,  $48 = 5 + 43$,  $50 = 7 + 43$,
    $52 = 5 + 47$,  $54 = 7 + 47$,  $56 = 3 + 53$,  $58 = 5 + 53$,
    $60 = 7 + 53$,  $62 = 3 + 59$,  $64 = 5 + 59$,  $66 = 7 + 59$,
    $68 = 7 + 61$,  $70 = 3 + 67$,  $72 = 5 + 67$,  $74 = 7 + 67$,
    $76 = 3 + 73$,  $78 = 5 + 73$,  $80 = 7 + 73$,  $82 = 3 + 79$,
    $84 = 5 + 79$,  $86 = 7 + 79$,  $88 = 5 + 83$,  $90 = 7 + 83$,
    $92 = 3 + 89$,  $94 = 5 + 89$,  $96 = 7 + 89$,  $98 = 19 + 79$.
3. Yes. For example, consider 94 and notice that $94 = 2 + 92$. If 92 were expressible as the sum of two odd primes, then 94 would be expressible as the sum of three primes. The idea generalizes for any even integer.
5. Only one. If $N = 2$, then $N^2 - 1 = 3$, a prime. For $N$ greater than 2, $N^2 - 1 = (N - 1)(N + 1)$ and neither factor $N - 1$ or $N + 1$ is 1; hence, $N^2 - 1$ has more than two distinct factors and it is not a prime number.
7. Yes.
9. Yes.


## Section 12.3, page 213

### Exercises I

1. $2 \times 2 \times 2 \times 3 \times 37$.
3. $2 \times 2 \times 3 \times 7 \times 11$.
5. $5 \times 11 \times 13$.
7. $2 \times 2 \times 3 \times 167$.
9. $2 \times 3 \times 3 \times 293$.

### Exercises II

1. If $a|b$ and $a|c$, then there exist integers $x$ and $y$ such that $ax = b$ and $ay = c$. Thus, $ax - ay = b - c$ and $a(x - y) = b - c$. Since $x - y$ is an integer, $a|(b - c)$.

Answers to Odd-Numbered Exercise Problems

3. (a) If $a|b$, then there exists an integer $x$ such that $ax = b$. Thus, $(ax)c = bc$ and $a(xc) = bc$. Since $xc$ is an integer, $a|bc$.
   (b) $4|(6)(10)$, but $4|6$ and $4|10$ are both false statements.
5. The set of positive integers can be separated into three disjoint sets as follows.

| Set I | Set II | Set III |
|---|---|---|
| 1 | 2 | 3 |
| 4 | 5 | 6 |
| 7 | 8 | 9 |
| 10 | 11 | 12 |
| .. | .. | .. |
| $3n - 2$ | $3n - 1$ | $3n$ |

Since $(3n)^2 = 3(3n^2)$, the square of every integer in Set III is in Set III. Since $(3n - 2)^2 = 9n^2 - 12n + 4 = 9n^2 - 12n + 6 - 2 = 3(3n^2 - 4n + 2) - 2$, the square of every integer in Set I is in Set I. However, since $(3n - 1)^2 = 9n^2 - 6n + 1 = 9n^2 - 6n + 3 - 2 = 3(3n^2 - 2n + 1) - 2$, the square of every integer in Set II is in Set I. Therefore, Set II contains no perfect squares.

7. The proof is completely similar to the one in the text.

## Section 12.4, page 219

**Exercises I**

1. 1.    3. 2.    5. 219.    7. 4.
9.
$$\begin{array}{r} 5 = q_1 \\ 168\overline{\smash{)}889} \\ 840 \quad\quad 3 = q_2 \\ \overline{49\overline{\smash{)}168}} \\ 147 \quad 2 = q_3 \\ \overline{21\overline{\smash{)}49}} \\ 42 \quad 3 = q_4 \\ G = 7\overline{\smash{)}21} \\ 21 \\ \overline{0} \end{array}$$

$u = 1 + q_2 q_3 = 7$ and $v = -q_1 - q_3 - q_1 q_2 q_3 = -37$. Thus,
$$7 = (889)(7) + (168)(-37).$$

## Exercises II

1. Since 1 is the greatest common factor of $a$ and $b$, there exist integers $u$ and $v$ such that $1 = au + bv$; thus, $c = cau + cbv$. Since $a|bc$ there exists an integer $x$ such that $ax = bc$. Substituting in the previous equation, $c = cau + axv$. Hence, $a(cu + xv) = c$. Since $(cu + xv)$ is an integer, $a|c$.

3.
$$
\begin{array}{r}
2 = q_1 \\
889 \overline{\smash{\big)}\,1946} \\
\underline{1778} \qquad 5 = q_2 \\
168 \overline{\smash{\big)}\,889} \\
\underline{840} \qquad 3 = q_3 \\
49 \overline{\smash{\big)}\,168} \\
\underline{147} \qquad 2 = q_4 \\
21 \overline{\smash{\big)}\,49} \\
\underline{42} \qquad 3 = q_5 \\
G = 7 \overline{\smash{\big)}\,21} \\
\underline{21} \\
0
\end{array}
$$

$7 = 1946(-5 - 2 - 30) + 889(1 + 10 + 4 + 6 + 60)$
$= 1946(-37) + 889(81)$
$= -72{,}002 + 72{,}009.$

5. Yes.

## Section 12.5, page 226

### Exercises I

1. 2.   3. 3.   5. 0.   7. 1.   9. 3.
11. (a) It is 0.   (b) It has none.
    (c) It is 4.   (d) It has none.
    (e) It is 5.   (f) It has none.
    (g) It is 6.   (h) It has none.

### Exercises II

1. $a \equiv b$, mod $m$, implies there is an integer $x$ such that $mx = a - b$. Multiplying by $c$, $(mx)c = ac - bc$ and $m(xc) = ac - bc$. Since $xc$ is an integer, $m$ is a factor of $ac - bc$ and $ac \equiv bc$, mod $m$.

Answers to Odd-Numbered Exercise Problems            279

3. Hint: Let $a = c$ and $b = d$, and see Theorem 5.
5. $a \equiv 0$, mod $m$, if and only if $a - 0 = a$ has $m$ as a factor by definition of congruence.
7. For any integer $x$, $x \equiv 0$, $x \equiv 1$, $x \equiv 2$, $x \equiv 3$, $x \equiv 4$, $x \equiv 5$, $x \equiv 6$, or $x \equiv 7$, mod 8. Thus, for any integer $x$, $x^2 \equiv 0$, $x^2 \equiv 1$, $x^2 \equiv 4$, $x^2 \equiv 9$, $x^2 \equiv 16$, $x^2 \equiv 25$, $x^2 \equiv 36$, or $x^2 \equiv 49$, mod 8. Since $9 \equiv 1$, $16 \equiv 0$, $25 \equiv 1$, $36 \equiv 4$, and $49 \equiv 1$, mod 8, we see that the square of any integer is congruent to 0, 1, or 4, mod 8.

## Section 12.6, page 233

**Exercises I**

1. 

| + | 0 | 1 | 2 | 3 | 4 | 5 | 6 | 7 |
|---|---|---|---|---|---|---|---|---|
| 0 | 0 | 1 | 2 | 3 | 4 | 5 | 6 | 7 |
| 1 | 1 | 2 | 3 | 4 | 5 | 6 | 7 | 10 |
| 2 | 2 | 3 | 4 | 5 | 6 | 7 | 10 | 11 |
| 3 | 3 | 4 | 5 | 6 | 7 | 10 | 11 | 12 |
| 4 | 4 | 5 | 6 | 7 | 10 | 11 | 12 | 13 |
| 5 | 5 | 6 | 7 | 10 | 11 | 12 | 13 | 14 |
| 6 | 6 | 7 | 10 | 11 | 12 | 13 | 14 | 15 |
| 7 | 7 | 10 | 11 | 12 | 13 | 14 | 15 | 16 |

| × | 0 | 1 | 2 | 3 | 4 | 5 | 6 | 7 |
|---|---|---|---|---|---|---|---|---|
| 0 | 0 | 0 | 0 | 0 | 0 | 0 | 0 | 0 |
| 1 | 0 | 1 | 2 | 3 | 4 | 5 | 6 | 7 |
| 2 | 0 | 2 | 4 | 6 | 10 | 12 | 14 | 16 |
| 3 | 0 | 3 | 6 | 11 | 14 | 17 | 22 | 25 |
| 4 | 0 | 4 | 10 | 14 | 20 | 24 | 30 | 34 |
| 5 | 0 | 5 | 12 | 17 | 24 | 31 | 36 | 43 |
| 6 | 0 | 6 | 14 | 22 | 30 | 36 | 44 | 52 |
| 7 | 0 | 7 | 16 | 25 | 34 | 43 | 52 | 61 |

3. (a) $\quad\begin{array}{r}1\,1\,1\,1_{\text{two}} = 15 \\ 1\,0\,1_{\text{two}} = \phantom{0}5 \\ \hline 1\,0\,1\,0\,0_{\text{two}} = 10\end{array}$   (b) $\begin{array}{r}1\,0\,1\,1\,0_{\text{two}} = 22 \\ 1\,0\,1\,1_{\text{two}} = 11 \\ \hline 1\,0\,1\,1_{\text{two}} = 11\end{array}$

(c) $\begin{array}{r}1\,0\,1\,1\,0\,1_{\text{two}} = 45 \\ 1\,0\,1\,1\,0_{\text{two}} = 22 \\ \hline 1\,0\,1\,1\,1_{\text{two}} = 23\end{array}$

5. (a) $\begin{array}{r}1\,0\,1_{\text{two}} \\ 1\,1_{\text{two}} \\ \hline 1\,0\,1 \\ 1\,0\,1 \\ \hline 1\,1\,1\,1_{\text{two}}\end{array}$   (b) $\begin{array}{r}1\,1\,1\,0\,0_{\text{two}} \\ 1\,1\,0\,1_{\text{two}} \\ \hline 1\,1\,1\,0\,0 \\ 1\,1\,1\,0\,0 \\ 1\,1\,1\,0\,0 \\ \hline 1\,0\,1\,1\,0\,1\,1\,0\,0_{\text{two}} \\ 1\,0\,1\,1_{\text{two}}\ \text{(add remainder)} \\ \hline 1\,0\,1\,1\,1\,0\,1\,1\,1_{\text{two}}\end{array}$

7. (a) $\begin{array}{r}1\,1\,0\,1_{\text{two}} = 13 \\ 1\,1_{\text{two}} = \phantom{0}3 \\ \hline 1\,0\,1\,0_{\text{two}} = 10\end{array}$   (b) $\begin{array}{r}1\,1\,0\,1\,0_{\text{two}} = 26 \\ 1\,1\,1\,1_{\text{two}} = 15 \\ \hline 1\,0\,1\,1_{\text{two}} = 11\end{array}$

(c) $\begin{array}{r}1\,0\,1\,1\,0\,0_{\text{two}} = 44 \\ 1\,0\,0\,1\,1_{\text{two}} = 19 \\ \hline 1\,1\,0\,0\,1_{\text{two}} = 25\end{array}$

9. (a) $8\,|\,7894$
$\phantom{0}8\,|\,\phantom{0}986 \quad r = 6$
$\phantom{0}8\,|\,\phantom{00}123 \quad r = 2$
$\phantom{0}8\,|\,\phantom{000}15 \quad r = 3$
$\phantom{0000000}1 \quad r = 7$
Thus, $7{,}894 = 17326_{\text{eight}}$

(b) $8\,|\,1354$
$\phantom{0}8\,|\,\phantom{0}169 \quad r = 2$
$\phantom{0}8\,|\,\phantom{00}21 \quad r = 1$
$\phantom{0000000}2 \quad r = 5$
Thus, $1{,}354 = 2512_{\text{eight}}$

(c) $8\,|\,243{,}567$
$\phantom{0}8\,|\,30{,}445 \quad r = 7$
$\phantom{0}8\,|\,\phantom{0}3{,}805 \quad r = 5$
$\phantom{0}8\,|\,\phantom{00}475 \quad r = 5$
$\phantom{0}8\,|\,\phantom{0000}59 \quad r = 3$
$\phantom{00000000}7 \quad r = 3$
Thus, $243{,}567 = 733557_{\text{eight}}$

(d) $8\,|\,24{,}640$
$\phantom{0}8\,|\,3{,}080 \quad r = 0$
$\phantom{0}8\,|\,\phantom{0}385 \quad r = 0$
$\phantom{0}8\,|\,\phantom{00}48 \quad r = 1$
$\phantom{00000000}6 \quad r = 0$
Thus, $24{,}640 = 60100_{\text{eight}}$

*Answers to Odd-Numbered Exercise Problems*

11. (a) $6744_{eight} = 3556$  (b) $7354_{eight} = 3820$
    $3522_{eight} = 1874$        $1716_{eight} = \phantom{0}974$
    $\overline{3222_{eight} = 1682}$   $\overline{5436_{eight} = 2846}$

(c) $7245_{eight} = 3749$
    $4766_{eight} = 2550$
    $\overline{2257_{eight} = 1199}$

13. (a) $6 \,|\, 8964$                     (b) $6 \,|\, 2345$
       $6 \,|\, 1497 \quad r = 2$             $6 \,|\, 390 \quad r = 5$
       $6 \,|\, 249 \quad r = 3$               $6 \,|\, 65 \quad r = 0$
       $6 \,|\, 41 \quad r = 3$                $6 \,|\, 10 \quad r = 5$
          $6 \quad r = 5$                    $1 \quad r = 4$
Thus, $8,964 = 65332_{six}$.      Thus, $2,345 = 14505_{six}$.

(c) $6 \,|\, 467,960$               (d) $6 \,|\, 3,600$
       $6 \,|\, 77,993 \quad r = 2$           $6 \,|\, 600 \quad r = 0$
       $6 \,|\, 12,998 \quad r = 5$           $6 \,|\, 100 \quad r = 0$
       $6 \,|\, 2,166 \quad r = 2$            $6 \,|\, 16 \quad r = 4$
       $6 \,|\, 361 \quad r = 0$                 $2 \quad r = 4$
       $6 \,|\, 60 \quad r = 1$       Thus, $3,600 = 24400_{six}$.
       $6 \,|\, 10 \quad r = 0$
          $1 \quad r = 4$
Thus, $467,960 = 14010252_{six}$.

### Exercises II

1. (a) A number is divisible by 2 if and only if the last digit is 0, 2, 4, or 6.
   (b) A number is divisible by 7 if and only if the sum of the digits is divisible by 7.
   (c) A number is divisible by $10_{eight}$ if and only if the last digit is 0.
   [The proofs are similar to the proofs for base ten.]

## Section 12.7, page 240

### Exercises I

1. $\frac{2}{7}$.      3. $\frac{218}{999}$.     5. 1.
7. $\frac{7}{33}$.     9. $\frac{35}{111}$.    11. $\frac{3578}{99}$.
13. $0.\overline{714285}\ldots$     15. $0.27\overline{27}\ldots$          17. $0.3750\overline{0}\ldots$

## Amount at Compound Interest $(1 + i)^n$

| Periods | Rate $i$ | | | | |
|---|---|---|---|---|---|
| $n$ | .0025 (¼%) | .004167 (1/24%) | .005 (½%) | .005833 (7/12%) | .0075 (¾%) |
| 1 | 1.0025 0000 | 1.0041 6667 | 1.0050 0000 | 1.0058 3333 | 1.0075 0000 |
| 2 | 1.0050 0625 | 1.0083 5069 | 1.0100 2500 | 1.0117 0069 | 1.0150 5625 |
| 3 | 1.0075 1877 | 1.0125 5216 | 1.0150 7513 | 1.0176 0228 | 1.0226 6917 |
| 4 | 1.0100 3756 | 1.0167 7112 | 1.0201 5050 | 1.0235 3830 | 1.0303 3919 |
| 5 | 1.0125 6266 | 1.0210 0767 | 1.0252 5125 | 1.0295 0894 | 1.0380 6673 |
| 6 | 1.0150 9406 | 1.0252 6187 | 1.0303 7751 | 1.0355 1440 | 1.0458 5224 |
| 7 | 1.0176 3180 | 1.0295 3379 | 1.0355 2940 | 1.0415 5490 | 1.0536 9613 |
| 8 | 1.0201 7588 | 1.0338 2352 | 1.0407 0704 | 1.0476 3064 | 1.0615 9885 |
| 9 | 1.0227 2632 | 1.0381 3111 | 1.0459 1058 | 1.0537 4182 | 1.0695 6084 |
| 10 | 1.0252 8313 | 1.0424 5666 | 1.0511 4013 | 1.0598 8865 | 1.0775 8255 |
| 11 | 1.0278 4634 | 1.0468 0023 | 1.0563 9583 | 1.0660 7133 | 1.0856 6441 |
| 12 | 1.0304 1596 | 1.0511 6190 | 1.0616 7781 | 1.0722 9008 | 1.0938 0690 |
| 13 | 1.0329 9200 | 1.0555 4174 | 1.0669 8620 | 1.0785 4511 | 1.1020 1045 |
| 14 | 1.0355 7448 | 1.0599 3983 | 1.0723 2113 | 1.0848 3662 | 1.1102 7553 |
| 15 | 1.0381 6341 | 1.0643 5625 | 1.0776 8274 | 1.0911 6483 | 1.1186 0259 |
| 16 | 1.0407 5882 | 1.0687 9106 | 1.0830 7115 | 1.0975 2996 | 1.1269 9211 |
| 17 | 1.0433 6072 | 1.0732 4436 | 1.0884 8651 | 1.1039 3222 | 1.1354 4455 |
| 18 | 1.0459 6912 | 1.0777 1621 | 1.0939 2894 | 1.1103 7182 | 1.1439 6039 |
| 19 | 1.0485 8404 | 1.0822 0670 | 1.0993 9858 | 1.1168 4899 | 1.1525 4009 |
| 20 | 1.0512 0550 | 1.0867 1589 | 1.1048 9558 | 1.1233 6395 | 1.1611 8414 |
| 21 | 1.0538 3352 | 1.0912 4387 | 1.1104 2006 | 1.1299 1690 | 1.1698 9302 |
| 22 | 1.0564 6810 | 1.0957 9072 | 1.1159 7216 | 1.1365 0808 | 1.1786 6722 |
| 23 | 1.0591 0927 | 1.1003 5652 | 1.1215 5202 | 1.1431 3771 | 1.1875 0723 |
| 24 | 1.0617 5704 | 1.1049 4134 | 1.1271 5978 | 1.1498 0602 | 1.1964 1353 |
| 25 | 1.0644 1144 | 1.1095 4526 | 1.1327 9558 | 1.1565 1322 | 1.2053 8663 |
| 26 | 1.0670 7247 | 1.1141 6836 | 1.1384 5955 | 1.1632 5955 | 1.2144 2703 |
| 27 | 1.0697 4015 | 1.1188 1073 | 1.1441 5185 | 1.1700 4523 | 1.2235 3523 |
| 28 | 1.0724 1450 | 1.1234 7244 | 1.1498 7261 | 1.1768 7049 | 1.2327 1175 |
| 29 | 1.0750 9553 | 1.1281 5358 | 1.1556 2197 | 1.1837 3557 | 1.2419 5709 |
| 30 | 1.0777 8327 | 1.1328 5422 | 1.1614 0008 | 1.1906 4069 | 1.2512 7176 |
| 31 | 1.0804 7773 | 1.1375 7444 | 1.1672 0708 | 1.1975 8610 | 1.2606 5630 |
| 32 | 1.0831 7892 | 1.1423 1434 | 1.1730 4312 | 1.2045 7202 | 1.2701 1122 |
| 33 | 1.0858 8687 | 1.1470 7398 | 1.1789 0833 | 1.2115 9869 | 1.2796 3706 |
| 34 | 1.0886 0159 | 1.1518 5346 | 1.1848 0288 | 1.2186 6634 | 1.2892 3434 |
| 35 | 1.0913 2309 | 1.1566 5284 | 1.1907 2689 | 1.2257 7523 | 1.2989 0359 |
| 36 | 1.0940 5140 | 1.1614 7223 | 1.1966 8052 | 1.2329 2559 | 1.3086 4537 |
| 37 | 1.0967 8653 | 1.1663 1170 | 1.2026 6393 | 1.2401 1765 | 1.3184 6021 |
| 38 | 1.0995 2850 | 1.1711 7133 | 1.2086 7725 | 1.2473 5167 | 1.3283 4866 |
| 39 | 1.1022 7732 | 1.1760 5121 | 1.2147 2063 | 1.2546 2789 | 1.3383 1128 |
| 40 | 1.1050 3301 | 1.1809 5142 | 1.2207 9424 | 1.2619 4655 | 1.3483 4861 |
| 41 | 1.1077 9559 | 1.1858 7206 | 1.2268 9821 | 1.2693 0791 | 1.3584 6123 |
| 42 | 1.1105 6508 | 1.1908 1319 | 1.2330 3270 | 1.2767 1220 | 1.3686 4969 |
| 43 | 1.1133 4149 | 1.1957 7491 | 1.2391 9786 | 1.2841 5969 | 1.3789 1456 |
| 44 | 1.1161 2485 | 1.2007 5731 | 1.2453 9385 | 1.2916 5062 | 1.3892 5642 |
| 45 | 1.1189 1516 | 1.2057 6046 | 1.2516 2082 | 1.2991 8525 | 1.3996 7584 |
| 46 | 1.1217 1245 | 1.2107 8446 | 1.2578 7892 | 1.3067 6383 | 1.4101 7341 |
| 47 | 1.1245 1673 | 1.2158 2940 | 1.2641 6832 | 1.3143 8662 | 1.4207 4971 |
| 48 | 1.1273 2802 | 1.2208 9536 | 1.2704 8916 | 1.3220 5388 | 1.4314 0533 |
| 49 | 1.1301 4634 | 1.2259 8242 | 1.2768 4161 | 1.3297 6586 | 1.4421 4087 |
| 50 | 1.1329 7171 | 1.2310 9068 | 1.2832 2581 | 1.3375 2283 | 1.4529 5693 |
| 51 | 1.1358 0414 | 1.2362 2022 | 1.2896 4194 | 1.3453 2504 | 1.4638 5411 |
| 52 | 1.1386 4365 | 1.2413 7114 | 1.2960 9015 | 1.3531 7277 | 1.4748 3301 |
| 53 | 1.1414 9026 | 1.2465 4352 | 1.3025 7060 | 1.3610 6628 | 1.4858 9426 |
| 54 | 1.1443 4398 | 1.2517 3745 | 1.3090 8346 | 1.3690 0583 | 1.4970 3847 |
| 55 | 1.1472 0484 | 1.2569 5302 | 1.3156 2887 | 1.3769 9170 | 1.5082 6626 |
| 56 | 1.1500 7285 | 1.2621 9033 | 1.3222 0702 | 1.3850 2415 | 1.5195 7825 |
| 57 | 1.1529 4804 | 1.2674 4946 | 1.3288 1805 | 1.3931 0346 | 1.5309 7509 |
| 58 | 1.1558 3041 | 1.2727 3050 | 1.3354 6214 | 1.4012 2990 | 1.5424 5740 |
| 59 | 1.1587 1998 | 1.2780 3354 | 1.3421 3946 | 1.4094 0374 | 1.5540 2583 |
| 60 | 1.1616 1678 | 1.2833 5868 | 1.3488 5015 | 1.4176 2526 | 1.5656 8103 |
| 61 | 1.1645 2082 | 1.2887 0601 | 1.3555 9440 | 1.4258 9474 | 1.5774 2363 |
| 62 | 1.1674 3213 | 1.2940 7561 | 1.3623 7238 | 1.4342 1246 | 1.5892 5431 |
| 63 | 1.1703 5071 | 1.2994 6760 | 1.3691 8424 | 1.4425 7870 | 1.6011 7372 |
| 64 | 1.1732 7658 | 1.3048 8204 | 1.3760 3016 | 1.4509 9374 | 1.6131 8252 |

| Periods | Rate $i$ | | | | |
|---|---|---|---|---|---|
| $n$ | .0025 ($\frac{1}{4}$%) | .004167 ($\frac{5}{12}$%) | .005 ($\frac{1}{2}$%) | .005833 ($\frac{7}{12}$%) | .0075 ($\frac{3}{4}$%) |
| 65 | 1.1762 0977 | 1.3103 1905 | 1.3829 1031 | 1.4594 5787 | 1.6252 8139 |
| 66 | 1.1791 5030 | 1.3157 7872 | 1.3898 2486 | 1.4679 7138 | 1.6374 7100 |
| 67 | 1.1820 9817 | 1.3212 6113 | 1.3967 7399 | 1.4765 3454 | 1.6497 5203 |
| 68 | 1.1850 5342 | 1.3267 6638 | 1.4037 5785 | 1.4851 4766 | 1.6621 2517 |
| 69 | 1.1880 1605 | 1.3322 9458 | 1.4107 7664 | 1.4938 1102 | 1.6745 9111 |
| 70 | 1.1909 8609 | 1.3378 4580 | 1.4178 3053 | 1.5025 2492 | 1.6871 5055 |
| 71 | 1.1939 6356 | 1.3434 2016 | 1.4249 1968 | 1.5112 8965 | 1.6998 0418 |
| 72 | 1.1969 4847 | 1.3490 1774 | 1.4320 4428 | 1.5201 0550 | 1.7125 5271 |
| 73 | 1.1999 4084 | 1.3546 3865 | 1.4392 0450 | 1.5289 7279 | 1.7253 9685 |
| 74 | 1.2029 4069 | 1.3602 8298 | 1.4464 0052 | 1.5378 9179 | 1.7383 3733 |
| 75 | 1.2059 4804 | 1.3659 5082 | 1.4536 3252 | 1.5468 6283 | 1.7513 7486 |
| 76 | 1.2089 6291 | 1.3716 4229 | 1.4609 0069 | 1.5558 8620 | 1.7645 1017 |
| 77 | 1.2119 8532 | 1.3773 5746 | 1.4682 0519 | 1.5649 6220 | 1.7777 4400 |
| 78 | 1.2150 1528 | 1.3830 9645 | 1.4755 4622 | 1.5740 9115 | 1.7910 7708 |
| 79 | 1.2180 5282 | 1.3888 5935 | 1.4829 2395 | 1.5832 7334 | 1.8045 1015 |
| 80 | 1.2210 9795 | 1.3946 4627 | 1.4903 3857 | 1.5925 0910 | 1.8180 4398 |
| 81 | 1.2241 5070 | 1.4004 5729 | 1.4977 9026 | 1.6017 9874 | 1.8316 7931 |
| 82 | 1.2272 1108 | 1.4062 9253 | 1.5052 7921 | 1.6111 4257 | 1.8454 1691 |
| 83 | 1.2302 7910 | 1.4121 5209 | 1.5128 0561 | 1.6205 4090 | 1.8592 5753 |
| 84 | 1.2333 5480 | 1.4180 3605 | 1.5203 6964 | 1.6299 9405 | 1.8732 0196 |
| 85 | 1.2364 3819 | 1.4239 4454 | 1.5279 7148 | 1.6395 0235 | 1.8872 5098 |
| 86 | 1.2395 2928 | 1.4298 7764 | 1.5356 1134 | 1.6490 6612 | 1.9014 0536 |
| 87 | 1.2426 2811 | 1.4358 3546 | 1.5432 8940 | 1.6586 8567 | 1.9156 6590 |
| 88 | 1.2457 3468 | 1.4418 1811 | 1.5510 0585 | 1.6683 6134 | 1.9300 3339 |
| 89 | 1.2488 4901 | 1.4478 2568 | 1.5587 6087 | 1.6780 9344 | 1.9445 0865 |
| 90 | 1.2519 7114 | 1.4538 5829 | 1.5665 5468 | 1.6878 8232 | 1.9590 9246 |
| 91 | 1.2551 0106 | 1.4599 1603 | 1.5743 8745 | 1.6977 2830 | 1.9737 8565 |
| 92 | 1.2582 3882 | 1.4659 9902 | 1.5822 5939 | 1.7076 3172 | 1.9885 8905 |
| 93 | 1.2613 8441 | 1.4721 0735 | 1.5901 7069 | 1.7175 9290 | 2.0035 0346 |
| 94 | 1.2645 3787 | 1.4782 4113 | 1.5981 2154 | 1.7276 1219 | 2.0185 2974 |
| 95 | 1.2676 9922 | 1.4844 0047 | 1.6061 1215 | 1.7376 8993 | 2.0336 6871 |
| 96 | 1.2708 6847 | 1.4905 8547 | 1.6141 4271 | 1.7478 2646 | 2.0489 2123 |
| 97 | 1.2740 4564 | 1.4967 9624 | 1.6222 1342 | 1.7580 2211 | 2.0642 8814 |
| 98 | 1.2772 3075 | 1.5030 3289 | 1.6303 2449 | 1.7682 7724 | 2.0797 7030 |
| 99 | 1.2804 2383 | 1.5092 9553 | 1.6384 7611 | 1.7785 9219 | 2.0953 6858 |
| 100 | 1.2836 2489 | 1.5155 8426 | 1.6466 6849 | 1.7889 6731 | 2.1110 8384 |

# Compound Interest Tables

| Periods | Rate $i$ | | | | |
|---|---|---|---|---|---|
| $n$ | .01 (1%) | .01125 (1⅛%) | .0125 (1¼%) | .015 (1½%) | .0175 (1¾%) |
| 1 | 1.0100 0000 | 1.0112 5000 | 1.0125 0000 | 1.0150 0000 | 1.0175 0000 |
| 2 | 1.0201 0000 | 1.0226 2656 | 1.0251 5625 | 1.0302 2500 | 1.0353 0625 |
| 3 | 1.0303 0100 | 1.0341 3111 | 1.0379 7070 | 1.0456 7838 | 1.0534 2411 |
| 4 | 1.0406 0401 | 1.0457 6509 | 1.0509 4534 | 1.0613 6355 | 1.0718 5903 |
| 5 | 1.0510 1005 | 1.0575 2994 | 1.0640 8215 | 1.0772 8400 | 1.0906 1656 |
| 6 | 1.0615 2015 | 1.0694 2716 | 1.0773 8318 | 1.0934 4326 | 1.1097 0235 |
| 7 | 1.0721 3535 | 1.0814 5821 | 1.0908 5047 | 1.1098 4491 | 1.1291 2215 |
| 8 | 1.0828 5671 | 1.0936 2462 | 1.1044 8610 | 1.1264 9259 | 1.1488 8178 |
| 9 | 1.0936 8527 | 1.1059 2789 | 1.1182 9218 | 1.1433 8998 | 1.1689 8721 |
| 10 | 1.1046 2213 | 1.1183 6958 | 1.1322 7083 | 1.1605 4083 | 1.1894 4449 |
| 11 | 1.1156 6835 | 1.1309 5124 | 1.1464 2422 | 1.1779 4894 | 1.2102 5977 |
| 12 | 1.1268 2503 | 1.1436 7444 | 1.1607 5452 | 1.1956 1817 | 1.2314 3931 |
| 13 | 1.1380 9328 | 1.1565 4078 | 1.1752 6395 | 1.2135 5244 | 1.2529 8950 |
| 14 | 1.1494 7421 | 1.1695 5186 | 1.1899 5475 | 1.2317 5573 | 1.2749 1682 |
| 15 | 1.1609 6896 | 1.1827 0932 | 1.2048 2918 | 1.2502 3207 | 1.2972 2786 |
| 16 | 1.1725 7864 | 1.1960 1480 | 1.2198 8955 | 1.2689 8555 | 1.3199 2935 |
| 17 | 1.1843 0443 | 1.2094 6997 | 1.2351 3817 | 1.2880 2033 | 1.3430 2811 |
| 18 | 1.1961 4748 | 1.2230 7650 | 1.2505 7739 | 1.3073 4064 | 1.3665 3111 |
| 19 | 1.2081 0895 | 1.2368 3611 | 1.2662 0961 | 1.3269 5075 | 1.3904 4540 |
| 20 | 1.2201 9004 | 1.2507 5052 | 1.2820 3723 | 1.3468 5501 | 1.4147 7820 |
| 21 | 1.2323 9194 | 1.2648 2146 | 1.2980 6270 | 1.3670 5783 | 1.4395 3681 |
| 22 | 1.2447 1586 | 1.2790 5071 | 1.3142 8848 | 1.3875 6370 | 1.4647 2871 |
| 23 | 1.2571 6302 | 1.2934 4003 | 1.3307 1709 | 1.4083 7715 | 1.4903 6146 |
| 24 | 1.2697 3465 | 1.3079 9123 | 1.3473 5105 | 1.4295 0281 | 1.5164 4279 |
| 25 | 1.2824 3200 | 1.3227 0613 | 1.3641 9294 | 1.4509 4535 | 1.5429 8054 |
| 26 | 1.2952 5631 | 1.3375 8657 | 1.3812 4535 | 1.4727 0953 | 1.5699 8269 |
| 27 | 1.3082 0588 | 1.3526 3442 | 1.3985 1092 | 1.4948 0018 | 1.5974 5739 |
| 28 | 1.3212 9097 | 1.3678 5156 | 1.4159 9230 | 1.5172 2218 | 1.6254 1290 |
| 29 | 1.3345 0388 | 1.3832 3989 | 1.4336 9221 | 1.5399 8051 | 1.6538 5762 |
| 30 | 1.3478 4892 | 1.3988 0134 | 1.4516 1336 | 1.5630 8022 | 1.6828 0013 |
| 31 | 1.3613 2740 | 1.4145 3785 | 1.4697 5853 | 1.5865 2642 | 1.7122 4913 |
| 32 | 1.3749 4068 | 1.4304 5140 | 1.4881 3051 | 1.6103 2432 | 1.7422 1349 |
| 33 | 1.3886 9009 | 1.4465 4398 | 1.5067 3214 | 1.6344 7915 | 1.7727 0223 |
| 34 | 1.4025 7699 | 1.4628 1760 | 1.5255 6629 | 1.6589 9637 | 1.8037 2452 |
| 35 | 1.4166 0276 | 1.4792 7430 | 1.5446 3587 | 1.6838 8132 | 1.8352 8970 |
| 36 | 1.4307 6878 | 1.4959 1613 | 1.5639 4382 | 1.7091 3954 | 1.8674 0727 |
| 37 | 1.4450 7647 | 1.5127 4519 | 1.5834 9312 | 1.7347 7663 | 1.9000 8689 |
| 38 | 1.4595 2724 | 1.5297 6357 | 1.6032 8678 | 1.7607 9828 | 1.9333 3841 |
| 39 | 1.4741 2251 | 1.5469 7341 | 1.6233 2787 | 1.7872 1025 | 1.9671 7184 |
| 40 | 1.4888 6373 | 1.5643 7687 | 1.6436 1946 | 1.8140 1841 | 2.0015 9734 |
| 41 | 1.5037 5237 | 1.5819 7611 | 1.6641 6471 | 1.8412 2868 | 2.0366 2530 |
| 42 | 1.5187 8989 | 1.5997 7334 | 1.6849 6673 | 1.8688 4712 | 2.0722 6624 |
| 43 | 1.5339 7779 | 1.6177 7079 | 1.7060 2885 | 1.8968 7982 | 2.1085 3090 |
| 44 | 1.5493 1757 | 1.6359 7071 | 1.7273 5421 | 1.9253 3302 | 2.1454 3019 |
| 45 | 1.5648 1075 | 1.6543 7538 | 1.7489 4614 | 1.9542 1301 | 2.1829 7522 |
| 46 | 1.5804 5885 | 1.6729 8710 | 1.7708 0797 | 1.9835 2621 | 2.2211 7728 |
| 47 | 1.5962 6344 | 1.6918 0821 | 1.7929 4306 | 2.0132 7910 | 2.2600 4789 |
| 48 | 1.6122 2608 | 1.7108 4105 | 1.8153 5485 | 2.0434 7829 | 2.2995 9872 |
| 49 | 1.6283 4834 | 1.7300 8801 | 1.8380 4679 | 2.0741 3046 | 2.3398 4170 |
| 50 | 1.6446 3182 | 1.7495 5150 | 1.8610 2237 | 2.1052 4242 | 2.3807 8893 |
| 51 | 1.6610 7814 | 1.7692 3395 | 1.8842 8515 | 2.1368 2106 | 2.4224 5274 |
| 52 | 1.6776 8892 | 1.7891 3784 | 1.9078 3872 | 2.1688 7337 | 2.4648 4566 |
| 53 | 1.6944 6581 | 1.8092 6564 | 1.9316 8670 | 2.2014 0647 | 2.5079 8046 |
| 54 | 1.7114 1047 | 1.8296 1988 | 1.9558 3279 | 2.2344 2757 | 2.5518 7012 |
| 55 | 1.7285 2457 | 1.8502 0310 | 1.9802 8070 | 2.2679 4398 | 2.5965 2785 |
| 56 | 1.7458 0982 | 1.8710 1788 | 2.0050 3420 | 2.3019 6314 | 2.6419 6708 |
| 57 | 1.7632 6792 | 1.8920 6684 | 2.0300 9713 | 2.3364 9259 | 2.6882 0151 |
| 58 | 1.7809 0060 | 1.9133 5259 | 2.0554 7335 | 2.3715 3998 | 2.7352 4503 |
| 59 | 1.7987 0960 | 1.9348 7780 | 2.0811 6676 | 2.4071 1308 | 2.7831 1182 |
| 60 | 1.8166 9670 | 1.9566 4518 | 2.1071 8135 | 2.4432 1978 | 2.8318 1628 |
| 61 | 1.8348 6367 | 1.9786 5744 | 2.1335 2111 | 2.4798 6807 | 2.8813 7306 |
| 62 | 1.8532 1230 | 2.0009 1733 | 2.1601 9013 | 2.5170 6609 | 2.9317 9709 |
| 63 | 1.8717 4443 | 2.0234 2765 | 2.1871 9250 | 2.5548 2208 | 2.9831 0354 |
| 64 | 1.8904 6187 | 2.0461 9121 | 2.2145 3241 | 2.5931 4442 | 3.0353 0785 |

| Periods | Rate $i$ | | | | |
|---|---|---|---|---|---|
| $n$ | .01 (1%) | .01125 (1⅛%) | .0125 (1¼%) | .015 (1½%) | .0175 (1¾%) |
| 65 | 1.9093 6649 | 2.0692 1087 | 2.2422 1407 | 2.6320 4158 | 3.0884 2574 |
| 66 | 1.9284 6015 | 2.0924 8949 | 2.2702 4174 | 2.6715 2221 | 3.1424 7319 |
| 67 | 1.9477 4475 | 2.1160 2999 | 2.2986 1976 | 2.7115 9504 | 3.1974 6647 |
| 68 | 1.9672 2220 | 2.1398 3533 | 2.3273 5251 | 2.7522 6896 | 3.2534 2213 |
| 69 | 1.9868 9442 | 2.1639 0848 | 2.3564 4442 | 2.7935 5300 | 3.3103 5702 |
| 70 | 2.0067 6337 | 2.1882 5245 | 2.3858 9997 | 2.8354 5629 | 3.3682 8827 |
| 71 | 2.0268 3100 | 2.2128 7029 | 2.4157 2372 | 2.8779 8814 | 3.4272 3331 |
| 72 | 2.0470 9931 | 2.2377 6508 | 2.4459 2027 | 2.9211 5796 | 3.4872 0990 |
| 73 | 2.0675 7031 | 2.2629 3994 | 2.4764 9427 | 2.9649 7533 | 3.5482 3607 |
| 74 | 2.0882 4601 | 2.2883 9801 | 2.5074 5045 | 3.0094 4996 | 3.6103 3020 |
| 75 | 2.1091 2847 | 2.3141 4249 | 2.5387 9358 | 3.0545 9171 | 3.6735 1098 |
| 76 | 2.1302 1975 | 2.3401 7659 | 2.5705 2850 | 3.1004 1059 | 3.7377 9742 |
| 77 | 2.1515 2195 | 2.3665 0358 | 2.6026 6011 | 3.1469 1674 | 3.8032 0888 |
| 78 | 2.1730 3717 | 2.3931 2675 | 2.6351 9336 | 3.1941 2050 | 3.8697 6503 |
| 79 | 2.1947 6754 | 2.4200 4942 | 2.6681 3327 | 3.2420 3230 | 3.9374 8592 |
| 80 | 2.2167 1522 | 2.4472 7498 | 2.7014 8494 | 3.2906 6279 | 4.0063 9192 |
| 81 | 2.2388 8237 | 2.4748 0682 | 2.7352 5350 | 3.3400 2273 | 4.0765 0378 |
| 82 | 2.2612 7119 | 2.5026 4840 | 2.7694 4417 | 3.3901 2307 | 4.1478 4260 |
| 83 | 2.2838 8390 | 2.5308 0319 | 2.8040 6222 | 3.4409 7492 | 4.2204 2984 |
| 84 | 2.3067 2274 | 2.5592 7473 | 2.8391 1300 | 3.4925 8954 | 4.2942 8737 |
| 85 | 2.3297 8997 | 2.5880 6657 | 2.8746 0191 | 3.5449 7838 | 4.3694 3740 |
| 86 | 2.3530 8787 | 2.6171 8232 | 2.9105 3444 | 3.5981 5306 | 4.4459 0255 |
| 87 | 2.3766 1875 | 2.6466 2562 | 2.9469 1612 | 3.6521 2535 | 4.5237 0584 |
| 88 | 2.4003 8494 | 2.6764 0016 | 2.9837 5257 | 3.7069 0723 | 4.6028 7070 |
| 89 | 2.4243 8879 | 2.7065 0966 | 3.0210 4948 | 3.7625 1084 | 4.6834 2093 |
| 90 | 2.4486 3267 | 2.7369 5789 | 3.0588 1260 | 3.8189 4851 | 4.7653 8080 |
| 91 | 2.4731 1900 | 2.7677 4867 | 3.0970 4775 | 3.8762 3273 | 4.8487 7496 |
| 92 | 2.4978 5019 | 2.7988 8584 | 3.1357 6085 | 3.9343 7622 | 4.9336 2853 |
| 93 | 2.5228 2869 | 2.8303 7331 | 3.1749 5786 | 3.9933 9187 | 5.0199 6703 |
| 94 | 2.5480 5698 | 2.8622 1501 | 3.2146 4483 | 4.0532 9275 | 5.1078 1645 |
| 95 | 2.5735 3755 | 2.8944 1492 | 3.2548 2789 | 4.1140 9214 | 5.1972 0324 |
| 96 | 2.5992 7293 | 2.9269 7709 | 3.2955 1324 | 4.1758 0352 | 5.2881 5429 |
| 97 | 2.6252 6565 | 2.9599 0559 | 3.3367 0716 | 4.2384 4057 | 5.3806 9699 |
| 98 | 2.6515 1831 | 2.9932 0452 | 3.3784 1600 | 4.3020 1718 | 5.4748 5919 |
| 99 | 2.6780 3349 | 3.0268 7807 | 3.4206 4620 | 4.3665 4744 | 5.5706 6923 |
| 100 | 2.7048 1383 | 3.0609 3045 | 3.4634 0427 | 4.4320 4565 | 5.6681 5594 |

## Compound Interest Tables

| Periods | Rate $i$ | | | | |
|---|---|---|---|---|---|
| $n$ | .02 (2%) | .0225 (2¼%) | .025 (2½%) | .0275 (2¾%) | .03 (3%) |
| 1 | 1.0200 0000 | 1.0225 0000 | 1.0250 0000 | 1.0275 0000 | 1.0300 0000 |
| 2 | 1.0404 0000 | 1.0455 0625 | 1.0506 2500 | 1.0557 5625 | 1.0609 0000 |
| 3 | 1.0612 0800 | 1.0690 3014 | 1.0768 9063 | 1.0847 8955 | 1.0927 2700 |
| 4 | 1.0824 3216 | 1.0930 8332 | 1.1038 1289 | 1.1146 2126 | 1.1255 0881 |
| 5 | 1.1040 8080 | 1.1176 7769 | 1.1314 0821 | 1.1452 7334 | 1.1592 7407 |
| 6 | 1.1261 6242 | 1.1428 2544 | 1.1596 9342 | 1.1767 6836 | 1.1940 5230 |
| 7 | 1.1486 8567 | 1.1685 3901 | 1.1886 8575 | 1.2091 2949 | 1.2298 7387 |
| 8 | 1.1716 5938 | 1.1948 3114 | 1.2184 0290 | 1.2423 8055 | 1.2667 7008 |
| 9 | 1.1950 9257 | 1.2217 1484 | 1.2488 6297 | 1.2765 4602 | 1.3047 7318 |
| 10 | 1.2189 9442 | 1.2492 0343 | 1.2800 8454 | 1.3116 5103 | 1.3439 1638 |
| 11 | 1.2433 7431 | 1.2773 1050 | 1.3120 8666 | 1.3477 2144 | 1.3842 3387 |
| 12 | 1.2682 4179 | 1.3060 4999 | 1.3448 8882 | 1.3847 8378 | 1.4257 6089 |
| 13 | 1.2936 0663 | 1.3354 3611 | 1.3785 1104 | 1.4228 6533 | 1.4685 3371 |
| 14 | 1.3194 7876 | 1.3654 8343 | 1.4129 7382 | 1.4619 9413 | 1.5125 8972 |
| 15 | 1.3458 6834 | 1.3962 0680 | 1.4482 9817 | 1.5021 9896 | 1.5579 6742 |
| 16 | 1.3727 8571 | 1.4276 2146 | 1.4845 0562 | 1.5435 0944 | 1.6047 0644 |
| 17 | 1.4002 4142 | 1.4597 4294 | 1.5216 1826 | 1.5859 5595 | 1.6528 4763 |
| 18 | 1.4282 4625 | 1.4925 8716 | 1.5596 5872 | 1.6295 6973 | 1.7024 3306 |
| 19 | 1.4568 1117 | 1.5261 7037 | 1.5986 5019 | 1.6743 8290 | 1.7535 0605 |
| 20 | 1.4859 4740 | 1.5605 0920 | 1.6386 1644 | 1.7204 2843 | 1.8061 1123 |
| 21 | 1.5156 6634 | 1.5956 2066 | 1.6795 8185 | 1.7677 4021 | 1.8602 9457 |
| 22 | 1.5459 7967 | 1.6315 2212 | 1.7215 7140 | 1.8163 5307 | 1.9161 0341 |
| 23 | 1.5768 9926 | 1.6682 3137 | 1.7646 1068 | 1.8663 0278 | 1.9735 8651 |
| 24 | 1.6084 3725 | 1.7057 6658 | 1.8087 2595 | 1.9176 2610 | 2.0327 9411 |
| 25 | 1.6406 0599 | 1.7441 4632 | 1.8539 4410 | 1.9703 6082 | 2.0937 7793 |
| 26 | 1.6734 1811 | 1.7833 8962 | 1.9002 9270 | 2.0245 4575 | 2.1565 9127 |
| 27 | 1.7068 8648 | 1.8235 1588 | 1.9478 0002 | 2.0802 2075 | 2.2212 8901 |
| 28 | 1.7410 2421 | 1.8645 4499 | 1.9964 9502 | 2.1374 2682 | 2.2879 2768 |
| 29 | 1.7758 4469 | 1.9064 9725 | 2.0464 0739 | 2.1962 0606 | 2.3565 6551 |
| 30 | 1.8113 6158 | 1.9493 9344 | 2.0975 6758 | 2.2566 0173 | 2.4272 6247 |
| 31 | 1.8475 8882 | 1.9932 5479 | 2.1500 0677 | 2.3186 5828 | 2.5000 8035 |
| 32 | 1.8845 4059 | 2.0381 0303 | 2.2037 5694 | 2.3824 2138 | 2.5750 8276 |
| 33 | 1.9222 3140 | 2.0839 6034 | 2.2588 5086 | 2.4479 3797 | 2.6523 3524 |
| 34 | 1.9606 7603 | 2.1308 4945 | 2.3153 2213 | 2.5152 5626 | 2.7319 0530 |
| 35 | 1.9998 8955 | 2.1787 9356 | 2.3732 0519 | 2.5844 2581 | 2.8138 6245 |
| 36 | 2.0398 8734 | 2.2278 1642 | 2.4325 3532 | 2.6554 9752 | 2.8982 7833 |
| 37 | 2.0806 8509 | 2.2779 4229 | 2.4933 4870 | 2.7285 2370 | 2.9852 2668 |
| 38 | 2.1222 9879 | 2.3291 9599 | 2.5556 8242 | 2.8035 5810 | 3.0747 8348 |
| 39 | 2.1647 4477 | 2.3816 0290 | 2.6195 7448 | 2.8806 5595 | 3.1670 2698 |
| 40 | 2.2080 3966 | 2.4351 8897 | 2.6850 6384 | 2.9598 7399 | 3.2620 3779 |
| 41 | 2.2522 0046 | 2.4899 8072 | 2.7521 9043 | 3.0412 7052 | 3.3598 9893 |
| 42 | 2.2972 4447 | 2.5460 0528 | 2.8209 9520 | 3.1249 0546 | 3.4606 9589 |
| 43 | 2.3431 8936 | 2.6032 9040 | 2.8915 2008 | 3.2108 4036 | 3.5645 1677 |
| 44 | 2.3900 5314 | 2.6618 6444 | 2.9638 0808 | 3.2991 3847 | 3.6714 5227 |
| 45 | 2.4378 5421 | 2.7217 5639 | 3.0379 0328 | 3.3898 6478 | 3.7815 9584 |
| 46 | 2.4866 1129 | 2.7829 9590 | 3.1138 5086 | 3.4830 8606 | 3.8950 4372 |
| 47 | 2.5363 4352 | 2.8456 1331 | 3.1916 9713 | 3.5788 7093 | 4.0118 9503 |
| 48 | 2.5870 7039 | 2.9096 3961 | 3.2714 8956 | 3.6772 8988 | 4.1322 5188 |
| 49 | 2.6388 1179 | 2.9751 0650 | 3.3532 7680 | 3.7784 1535 | 4.2562 1944 |
| 50 | 2.6915 8803 | 3.0420 4640 | 3.4371 0872 | 3.8823 2177 | 4.3839 0602 |
| 51 | 2.7454 1979 | 3.1104 9244 | 3.5230 3644 | 3.9890 8562 | 4.5154 2320 |
| 52 | 2.8003 2819 | 3.1804 7852 | 3.6111 1235 | 4.0987 8547 | 4.6508 8590 |
| 53 | 2.8563 3475 | 3.2520 3929 | 3.7013 9016 | 4.2115 0208 | 4.7904 1247 |
| 54 | 2.9134 6144 | 3.3252 1017 | 3.7939 2491 | 4.3273 1838 | 4.9341 2485 |
| 55 | 2.9717 3067 | 3.4000 2740 | 3.8887 7303 | 4.4463 1964 | 5.0821 4859 |
| 56 | 3.0311 6529 | 3.4765 2802 | 3.9859 9236 | 4.5685 9343 | 5.2346 1305 |
| 57 | 3.0917 8859 | 3.5547 4990 | 4.0856 4217 | 4.6942 2975 | 5.3916 5144 |
| 58 | 3.1536 2436 | 3.6347 3177 | 4.1877 8322 | 4.8233 2107 | 5.5534 0098 |
| 59 | 3.2166 9685 | 3.7165 1324 | 4.2924 7780 | 4.9559 6239 | 5.7200 0301 |
| 60 | 3.2810 3079 | 3.8001 3479 | 4.3997 8975 | 5.0922 5136 | 5.8916 0310 |
| 61 | 3.3466 5140 | 3.8856 3782 | 4.5097 8449 | 5.2322 8827 | 6.0683 5120 |
| 62 | 3.4135 8443 | 3.9730 6467 | 4.6225 2910 | 5.3761 7620 | 6.2504 0173 |
| 63 | 3.4818 5612 | 4.0624 5862 | 4.7380 9233 | 5.5240 2105 | 6.4379 1379 |
| 64 | 3.5514 9324 | 4.1538 6394 | 4.8565 4464 | 5.6759 3162 | 6.6310 5120 |

| Periods | Rate $i$ | | | | |
|---|---|---|---|---|---|
| $n$ | .02 (2%) | .0225 (2¼%) | .025 (2½%) | .0275 (2¾%) | .03 (3%) |
| 65 | 3.6225 2311 | 4.2473 2588 | 4.9779 5826 | 5.8320 1974 | 6.8299 8273 |
| 66 | 3.6949 7357 | 4.3428 9071 | 5.1024 0721 | 5.9924 0029 | 7.0348 8222 |
| 67 | 3.7688 7304 | 4.4406 0576 | 5.2299 6739 | 6.1571 9130 | 7.2459 2868 |
| 68 | 3.8442 5050 | 4.5405 1939 | 5.3607 1658 | 6.3265 1406 | 7.4633 0654 |
| 69 | 3.9211 3551 | 4.6426 8107 | 5.4947 3449 | 6.5004 9319 | 7.6872 0574 |
| 70 | 3.9995 5822 | 4.7471 4140 | 5.6321 0286 | 6.6792 5676 | 7.9178 2191 |
| 71 | 4.0795 4939 | 4.8539 5208 | 5.7729 0543 | 6.8629 3632 | 8.1553 5657 |
| 72 | 4.1611 4038 | 4.9631 6600 | 5.9172 2806 | 7.0516 6706 | 8.4000 1727 |
| 73 | 4.2443 6318 | 5.0748 3723 | 6.0651 5876 | 7.2455 8791 | 8.6520 1778 |
| 74 | 4.3292 5045 | 5.1890 2107 | 6.2167 8773 | 7.4448 4158 | 8.9115 7832 |
| 75 | 4.4158 3546 | 5.3057 7405 | 6.3722 0743 | 7.6495 7472 | 9.1789 2567 |
| 76 | 4.5041 5216 | 5.4251 5396 | 6.5315 1261 | 7.8599 3802 | 9.4542 9344 |
| 77 | 4.5942 3521 | 5.5472 1993 | 6.6948 0043 | 8.0760 8632 | 9.7379 2224 |
| 78 | 4.6861 1991 | 5.6720 3237 | 6.8621 7044 | 8.2981 7869 | 10.0300 5991 |
| 79 | 4.7798 4231 | 5.7996 5310 | 7.0337 2470 | 8.5263 7861 | 10.3309 6171 |
| 80 | 4.8754 3916 | 5.9301 4530 | 7.2095 6782 | 8.7608 5402 | 10.6408 9056 |
| 81 | 4.9729 4794 | 6.0635 7357 | 7.3898 07C1 | 9.0017 7751 | 10.9601 1727 |
| 82 | 5.0724 0690 | 6.2000 0397 | 7.5745 5219 | 9.2493 2639 | 11.2889 2079 |
| 83 | 5.1738 5504 | 6.3395 0406 | 7.7639 1599 | 9.5036 8286 | 11.6275 8842 |
| 84 | 5.2773 3214 | 6.4821 4290 | 7.9580 1389 | 9.7650 3414 | 11.9764 1607 |
| 85 | 5.3828 7878 | 6.6279 9112 | 8.1569 6424 | 10.0335 7258 | 12.3357 0855 |
| 86 | 5.4905 3636 | 6.7771 2092 | 8.3608 8834 | 10.3094 9583 | 12.7057 7981 |
| 87 | 5.6003 4708 | 6.9296 0614 | 8.5699 1055 | 10.5930 0696 | 13.0869 5320 |
| 88 | 5.7123 5402 | 7.0855 2228 | 8.7841 5832 | 10.8843 1465 | 13.4795 6180 |
| 89 | 5.8266 0110 | 7.2449 4653 | 9.0037 6228 | 11.1836 3331 | 13.8839 4865 |
| 90 | 5.9431 3313 | 7.4079 5782 | 9.2288 5633 | 11.4911 8322 | 14.3004 6711 |
| 91 | 6.0619 9579 | 7.5746 3688 | 9.4595 7774 | 11.8071 9076 | 14.7294 8112 |
| 92 | 6.1832 3570 | 7.7450 6621 | 9.6960 6718 | 12.1318 8851 | 15.1713 6556 |
| 93 | 6.3069 0042 | 7.9193 3020 | ·9.9384 6886 | 12.4655 1544 | 15.6265 0652 |
| 94 | 6.4330 3843 | 8.0975 1512 | 10.1869 3058 | 12.8083 1711 | 16.0953 0172 |
| 95 | 6.5616 9920 | 8.2797 0921 | 10.4416 0385 | 13.1605 4584 | 16.5781 6077 |
| 96 | 6.6929 3318 | 8.4660 0267 | 10.7026 4395 | 13.5224 6085 | 17.0755 0559 |
| 97 | 6.8267 9184 | 8.6564 8773 | 10.9702 1004 | 13.8943 2852 | 17.5877 7076 |
| 98 | 6.9633 2768 | 8.8512 5871 | 11.2444 6530 | 14.2764 2255 | 18.1154 0388 |
| 99 | 7.1025 9423 | 9.0504 1203 | 11.5255 7693 | 14.6690 2417 | 18.6588 6600 |
| 100 | 7.2446 4612 | 9.2540 4630 | 11.8137 1635 | 15.0724 2234 | 19.2186 3198 |

## Compound Interest Tables

| Periods | Rate $i$ | | | | |
|---|---|---|---|---|---|
| $n$ | .035 (3½%) | .04 (4%) | .045 (4½%) | .05 (5%) | .055 (5½%) |
| 1 | 1.0350 0000 | 1.0400 0000 | 1.0450 0000 | 1.0500 0000 | 1.0550 0000 |
| 2 | 1.0712 2500 | 1.0816 0000 | 1.0920 2500 | 1.1025 0000 | 1.1130 2500 |
| 3 | 1.1087 1788 | 1.1248 6400 | 1.1411 6613 | 1.1576 2500 | 1.1742 4138 |
| 4 | 1.1475 2300 | 1.1698 5856 | 1.1925 1860 | 1.2155 0625 | 1.2388 2465 |
| 5 | 1.1876 8631 | 1.2166 5290 | 1.2461 8194 | 1.2762 8156 | 1.3069 6001 |
| 6 | 1.2292 5533 | 1.2653 1902 | 1.3022 6012 | 1.3400 9564 | 1.3788 4281 |
| 7 | 1.2722 7926 | 1.3159 3178 | 1.3608 6183 | 1.4071 0042 | 1.4546 7916 |
| 8 | 1.3168 0904 | 1.3685 6905 | 1.4221 0061 | 1.4774 5544 | 1.5346 8651 |
| 9 | 1.3628 9735 | 1.4233 1181 | 1.4860 9514 | 1.5513 2822 | 1.6190 9427 |
| 10 | 1.4105 9876 | 1.4802 4428 | 1.5529 6942 | 1.6288 9463 | 1.7081 4446 |
| 11 | 1.4599 6972 | 1.5394 5406 | 1.6228 5305 | 1.7103 3936 | 1.8020 9240 |
| 12 | 1.5110 6866 | 1.6010 3222 | 1.6958 8143 | 1.7958 5633 | 1.9012 0749 |
| 13 | 1.5639 5606 | 1.6650 7351 | 1.7721 9610 | 1.8856 4914 | 2.0057 7390 |
| 14 | 1.6186 9452 | 1.7316 7645 | 1.8519 4492 | 1.9799 3160 | 2.1160 9146 |
| 15 | 1.6753 4883 | 1.8009 4351 | 1.9352 8244 | 2.0789 2818 | 2.2324 7649 |
| 16 | 1.7339 8604 | 1.8729 8125 | 2.0223 7015 | 2.1828 7459 | 2.3552 6270 |
| 17 | 1.7946 7555 | 1.9479 0050 | 2.1133 7681 | 2.2920 1832 | 2.4848 0215 |
| 18 | 1.8574 8920 | 2.0258 1652 | 2.2084 7877 | 2.4066 1923 | 2.6214 6627 |
| 19 | 1.9225 0132 | 2.1068 4918 | 2.3078 6031 | 2.5269 5020 | 2.7656 4691 |
| 20 | 1.9897 8886 | 2.1911 2314 | 2.4117 1402 | 2.6532 9771 | 2.9177 5749 |
| 21 | 2.0594 3147 | 2.2787 6807 | 2.5202 4116 | 2.7859 6259 | 3.0782 3415 |
| 22 | 2.1315 1158 | 2.3699 1879 | 2.6336 5201 | 2.9252 6072 | 3.2475 3703 |
| 23 | 2.2061 1448 | 2.4647 1554 | 2.7521 6635 | 3.0715 2376 | 3.4261 5157 |
| 24 | 2.2833 2849 | 2.5633 0416 | 2.8760 1383 | 3.2250 9994 | 3.6145 8990 |
| 25 | 2.3632 4498 | 2.6658 3633 | 3.0054 3446 | 3.3863 5494 | 3.8133 9235 |
| 26 | 2.4459 5856 | 2.7724 6978 | 3.1406 7901 | 3.5556 7269 | 4.0231 2893 |
| 27 | 2.5315 6711 | 2.8833 6858 | 3.2820 0956 | 3.7334 5632 | 4.2444 0102 |
| 28 | 2.6201 7196 | 2.9987 0332 | 3.4296 9999 | 3.9201 2914 | 4.4778 4307 |
| 29 | 2.7118 7798 | 3.1186 5145 | 3.5840 3649 | 4.1161 3560 | 4.7241 2444 |
| 30 | 2.8067 9370 | 3.2433 9751 | 3.7453 1813 | 4.3219 4238 | 4.9839 5129 |
| 31 | 2.9050 3148 | 3.3731 3341 | 3.9138 5745 | 4.5380 3949 | 5.2580 6861 |
| 32 | 3.0067 0759 | 3.5080 5875 | 4.0899 8104 | 4.7649 4147 | 5.5472 6238 |
| 33 | 3.1119 4235 | 3.6483 8110 | 4.2740 3018 | 5.0031 8854 | 5.8523 6181 |
| 34 | 3.2208 6033 | 3.7943 1634 | 4.4663 6154 | 5.2533 4797 | 6.1742 4171 |
| 35 | 3.3335 9045 | 3.9460 8899 | 4.6673 4781 | 5.5160 1537 | 6.5138 2501 |
| 36 | 3.4502 6611 | 4.1039 3255 | 4.8773 7846 | 5.7918 1614 | 6.8720 8538 |
| 37 | 3.5710 2543 | 4.2680 8986 | 5.0968 6049 | 6.0814 0694 | 7.2500 5008 |
| 38 | 3.6960 1132 | 4.4388 1345 | 5.3262 1921 | 6.3854 7729 | 7.6488 0283 |
| 39 | 3.8253 7171 | 4.6163 6599 | 5.5658 9908 | 6.7047 5115 | 8.0694 8699 |
| 40 | 3.9592 5972 | 4.8010 2063 | 5.8163 6454 | 7.0399 8871 | 8.5133 0877 |
| 41 | 4.0978 3381 | 4.9930 6145 | 6.0781 0094 | 7.3919 8815 | 8.9815 4076 |
| 42 | 4.2412 5799 | 5.1927 8391 | 6.3516 1548 | 7.7615 8756 | 9.4755 2550 |
| 43 | 4.3897 0202 | 5.4004 9527 | 6.6374 3818 | 8.1496 6693 | 9.9966 7940 |
| 44 | 4.5433 4160 | 5.6165 1508 | 6.9361 2290 | 8.5571 5028 | 10.5464 9677 |
| 45 | 4.7023 5855 | 5.8411 7568 | 7.2482 4843 | 8.9850 0779 | 11.1265 5409 |
| 46 | 4.8669 4110 | 6.0748 2271 | 7.5744 1961 | 9.4342 5818 | 11.7385 1456 |
| 47 | 5.0372 8404 | 6.3178 1562 | 7.9152 6849 | 9.9059 7109 | 12.3841 3287 |
| 48 | 5.2135 8898 | 6.5705 2824 | 8.2714 5557 | 10.4012 6965 | 13.0652 6017 |
| 49 | 5.3960 6459 | 6.8333 4937 | 8.6436 7107 | 10.9213 3313 | 13.7838 4948 |
| 50 | 5.5849 2686 | 7.1066 8335 | 9.0326 3627 | 11.4673 9979 | 14.5419 6120 |

## Compound Interest Tables

| Periods | Rate $i$ | | | | |
|---|---|---|---|---|---|
| $n$ | .06 (6%) | .065 (6½%) | .07 (7%) | .075 (7½%) | .08 (8%) |
| 1 | 1.0600 0000 | 1.0650 0000 | 1.0700 0000 | 1.0750 0000 | 1.0800 0000 |
| 2 | 1.1236 0000 | 1.1342 2500 | 1.1449 0000 | 1.1556 2500 | 1.1664 0000 |
| 3 | 1.1910 1600 | 1.2079 4963 | 1.2250 4300 | 1.2422 9688 | 1.2597 1200 |
| 4 | 1.2624 7696 | 1.2864 6635 | 1.3107 9601 | 1.3354 6914 | 1.3604 8896 |
| 5 | 1.3382 2558 | 1.3700 8666 | 1.4025 5173 | 1.4356 2933 | 1.4693 2808 |
| 6 | 1.4185 1911 | 1.4591 4230 | 1.5007 3035 | 1.5433 0153 | 1.5868 7432 |
| 7 | 1.5036 3026 | 1.5539 8655 | 1.6057 8148 | 1.6590 4914 | 1.7138 2427 |
| 8 | 1.5938 4807 | 1.6549 9567 | 1.7181 8618 | 1.7834 7783 | 1.8509 3021 |
| 9 | 1.6894 7896 | 1.7625 7039 | 1.8384 5921 | 1.9172 3866 | 1.9990 0463 |
| 10 | 1.7908 4770 | 1.8771 3747 | 1.9671 5136 | 2.0610 3156 | 2.1589 2500 |
| 11 | 1.8982 9856 | 1.9991 5140 | 2.1048 5195 | 2.2156 0893 | 2.3316 3900 |
| 12 | 2.0121 9647 | 2.1290 9624 | 2.2521 9159 | 2.3817 7960 | 2.5181 7012 |
| 13 | 2.1329 2826 | 2.2674 8750 | 2.4098 4500 | 2.5604 1307 | 2.7196 2373 |
| 14 | 2.2609 0396 | 2.4148 7418 | 2.5785 3415 | 2.7524 4405 | 2.9371 9362 |
| 15 | 2.3965 5819 | 2.5718 4101 | 2.7590 3154 | 2.9588 7735 | 3.1721 6911 |
| 16 | 2.5403 5168 | 2.7390 1067 | 2.9521 6375 | 3.1807 9315 | 3.4259 4264 |
| 17 | 2.6927 7279 | 2.9170 4637 | 3.1588 1521 | 3.4193 5264 | 3.7000 1805 |
| 18 | 2.8543 3915 | 3.1066 5438 | 3.3799 3228 | 3.6758 0409 | 3.9960 1950 |
| 19 | 3.0255 9950 | 3.3085 8691 | 3.6165 2754 | 3.9514 8940 | 4.3157 0106 |
| 20 | 3.2071 3547 | 3.5236 4506 | 3.8696 8446 | 4.2478 5110 | 4.6609 5714 |
| 21 | 3.3995 6360 | 3.7526 8199 | 4.1405 6237 | 4.5664 3993 | 5.0338 3372 |
| 22 | 3.6035 3742 | 3.9966 0632 | 4.4304 0174 | 4.9089 2293 | 5.4365 4041 |
| 23 | 3.8197 4966 | 4.2563 8573 | 4.7405 2986 | 5.2770 9215 | 5.8714 6365 |
| 24 | 4.0489 3464 | 4.5330 5081 | 5.0723 6695 | 5.6728 7406 | 6.3411 8074 |
| 25 | 4.2918 7072 | 4.8276 9911 | 5.4274 3264 | 6.0983 3961 | 6.8484 7520 |
| 26 | 4.5493 8296 | 5.1414 9955 | 5.8073 5292 | 6.5557 1508 | 7.3963 5321 |
| 27 | 4.8223 4594 | 5.4756 9702 | 6.2138 6763 | 7.0473 9371 | 7.9880 6147 |
| 28 | 5.1116 8670 | 5.8316 1733 | 6.6488 3836 | 7.5759 4824 | 8.6271 0639 |
| 29 | 5.4183 8790 | 6.2106 7245 | 7.1142 5705 | 8.1441 4436 | 9.3172 7490 |
| 30 | 5.7434 9117 | 6.6143 6616 | 7.6122 5504 | 8.7549 5519 | 10.0626 5689 |
| 31 | 6.0881 0064 | 7.0442 9996 | 8.1451 1290 | 9.4115 7683 | 10.8676 6944 |
| 32 | 6.4533 8668 | 7.5021 7946 | 8.7152 7080 | 10.1174 4509 | 11.7370 8300 |
| 33 | 6.8405 8988 | 7.9898 2113 | 9.3253 3975 | 10.8762 5347 | 12.6760 4964 |
| 34 | 7.2510 2528 | 8.5091 5950 | 9.9781 1354 | 11.6919 7248 | 13.6901 3361 |
| 35 | 7.6860 8679 | 9.0622 5487 | 10.6765 8148 | 12.5688 7042 | 14.7853 4429 |
| 36 | 8.1472 5200 | 9.6513 0143 | 11.4239 4219 | 13.5115 3570 | 15.9681 7184 |
| 37 | 8.6360 8712 | 10.2786 3603 | 12.2236 1814 | 14.5249 0088 | 17.2456 2558 |
| 38 | 9.1542 5235 | 10.9467 4737 | 13.0792 7141 | 15.6142 6844 | 18.6252 7563 |
| 39 | 9.7035 0749 | 11.6582 8595 | 13.9948 2041 | 16.7853 3858 | 20.1152 9768 |
| 40 | 10.2857 1794 | 12.4160 7453 | 14.9744 5784 | 18.0442 3897 | 21.7245 2150 |
| 41 | 10.9028 6101 | 13.2231 1938 | 16.0226 6989 | 19.3975 5689 | 23.4624 8322 |
| 42 | 11.5570 3267 | 14.0826 2214 | 17.1442 5678 | 20.8523 7366 | 25.3394 8187 |
| 43 | 12.2504 5463 | 14.9979 9258 | 18.3443 5475 | 22.4163 0168 | 27.3666 4042 |
| 44 | 12.9854 8191 | 15.9728 6209 | 19.6284 5959 | 24.0975 2431 | 29.5559 7166 |
| 45 | 13.7646 1083 | 17.0110 9813 | 21.0024 5176 | 25.9048 3863 | 31.9204 4939 |
| 46 | 14.5904 8748 | 18.1168 1951 | 22.4726 2338 | 27.8477 0153 | 34.4740 8534 |
| 47 | 15.4659 1673 | 19.2944 1278 | 24.0457 0702 | 29.9362 7915 | 37.2320 1217 |
| 48 | 16.3938 7173 | 20.5485 4961 | 25.7289 0651 | 32.1815 0008 | 40.2105 7314 |
| 49 | 17.3775 0403 | 21.8842 0533 | 27.5299 2997 | 34.5951 1259 | 43.4274 1899 |
| 50 | 18.4201 5427 | 23.3066 7868 | 29.4570 2506 | 37.1897 4603 | 46.9016 1251 |

The tables in this section are taken from the *C. R. C. Standard Mathematical Tables*, 17th edition. (The Chemical Rubber Publishing Company, Cleveland, Ohio.) Permission to reproduce these tables was granted by The Chemical Rubber Publishing Company.

# INDEX

Acute angle, 184
Addends, 21
Addition:
   associative property for, 23, 118
   base eight notation, 227–230
   base two notation, 230–233
   carrying in, 29
   closure property for, 21, 24, 37, 116
   commutative property for, 22, 117
   with decimals, 127–129
   with fractions, 115–120
   of integers, 154–157
   with mixed numerals, 120–124
   object-set interpretation, 21
   precedence of multiplication over, 53
   of rational numbers, 115–128
   subtraction as inverse operation of, 39
   of whole numbers, 21–35
Additive identity, 24
Additive inverses, 153
Algebraic operations, 243
Angle, acute, 184
Applications, 165–201
Approximation rationals with decimals, 129–135
Area, 137, 170–176
Arithmetic, basic features of, 3
Arithmetic mean, 193
Arithmetic sequence, 189–196
Associative property:
   for addition of rational numbers, 118
   for addition of whole numbers, 24
   for multiplication of rational numbers, 139
   for multiplication of whole numbers, 52
   for set intersection, 14
   for set union, 13–14
Average, 193

Base eight, 227–230
Base number, 4
Base ten, 4–11
Base twelve, 9–10
Base two, 230–234
Binary operation:
   for sets, 12
   of addition, 21
Binary system, 230–234
Borrowing, in subtraction, 41

Cardinal numbers, 3–4
Carrying in addition, 29
Casting-out-nines check, 33–34, 42, 58
Center of circle, 174
Center of sphere, 178
Centimeter, 168
Circle, 174
   circumference of, 174
Closure property:
   for addition, 21, 24, 116
   for division, 141
   for multiplication, 51
   for numbers, 37
   for subtraction, 37–38
Common difference, 192
Common ratio, 198
Commutative property:
   for addition of rational numbers, 117
   for addition of whole numbers, 22
   for multiplication of rational numbers, 139
   for multiplication of whole numbers, 51
   for set intersection, 14
   for set union, 13
Complement of a set, 14–15
Complex decimals, 148–150
Complex fractions, 147–151
Complex numbers, 243

Composite numbers, 84, 211–212
Compound interest, 200
Congruence of numbers, 220–227
Correspondence, one-to-one, 15
Cylinder, 179

Decimal fraction, 126
Decimal notation, 124–129
Decimal point, 125
Decimals:
  addition with, 127–129
  complex, 148–150
  division with, 144–147
  finite, 126
  infinite, 237
  multiplication with, 143–150
  $n$-place, 126
  for rational approximation, 129–134
  repeating, 237
  simple, 125
  terminating, 126
Decimeter, 168
Degree, 184
Denominator, 104
Difference:
  definition, 37
  of rational numbers, 118
  of whole numbers, 37–40
Digits, 7
Disjoint sets, 13
Distributive property, 53
Dividend, 65
Division:
  algorithm, 73–75
  closure for, 65
  with decimals, 144–146
  definition, 65
  incomplete, 71–75
  long, 73
  of rationals, 140–143
  by zero, 66, 141
Divisibility test, 65–70, 203–207
Divisor, 65
Dry measure, 181
Duodecimal system, 10

Empty set, 13
English system of measure, 167
Equality:
  properties, 22
  of rational numbers, 108
Eratosthenes, Sieve of, 84
Equivalent sets, 15
Euclidean algorithm, 94–98, 214–220
Even natural numbers, 66
Exponents, 125

Factor:
  definition, 59
  greatest common factor, 98
  prime, 88, 210–214
Factor tree, 87–88
Feet, 167
Finite decimals, 126
Fractions:
  addition with, 115–120
  complex, 147–151
  decimal, 126–129
  denominator of, 104
  division with, 140–143
  improper, 105
  least common denominator, 110
  multiplication with, 137–140
  notation, 102–113
  numerator of, 104
  proper, 104
  reduced to lowest terms, 109
Fundamental Theorem of Arithmetic, 87–89, 210–214

Geometric series, 196–201
Greater than, 43, 109
Greatest common factor, 89, 96
Greatest element, 44

Halve-double-sum technique, 76–79
Headline of a table, 25
Hexagon, 170
Hindu-Arabic system, 8

Identity, additive, 24
Improper fraction, 105
Inches, 167
Incomplete division, 71
Incomplete quotient, 71
Inequality, 43, 110, 161
Infinite decimals, 235–240
Infinite geometric series, 236
Infinitude of primes, 207, 208
Initial point, 103
Integers, 153–162
Interest, 200
Intersection of sets, 12
Inverse:
  additive, 39
  multiplicative, 139
Irrational numbers, 240–243
Isomorphic, 153

Lame's Theorem, 97
Least common denominator, 110
Least common multiple, 91
Least element, 44
Less than, 43, 110
Line, 103

# Index

Linear measure, 166–170
Liquid measure, 181
Long division, 73
Lowest terms, 109

Main diagonal of a table, 25
Mean, arithmetic, 193
Measure:
  of area, 173
  dry, 181
  English system of, 167
  linear, 166–170
  liquid, 181
Meter, 168
Metric system of linear measure, 166–170
Mile, 167
Millimeter, 168
Minuend, 37
Mixed numeral notation, 120–124
Modulo $m$, 220–227
Multiple, 53
Multiplicand, 57
Multiplication:
  associative property for, 52, 139
  base eight notation, 227–230
  base two notation, 230–234
  closure property for, 51
  commutative property for, 51, 139
  with decimals, 143–150
  with fractions, 137–140
  of integers, 158–160
  with mixed numerals, 141
  precedence over division, 53
  of rational numbers, 138
  techniques, 55–59, 76–79
  of whole numbers, 49
Multiplicative identity, 51
Multiplicative inverses, 139
Multiplier, 57

Natural numbers, 5
Negative integers, 153
Negative rationals, 156
Notation:
  base eight, 227
  base ten, 4–11
  binary, 230–234
  decimal, 124–129
  fractional, 102–113
  Hindu-Arabic, 8
  mixed numeral, 120–124
  Roman, 8
$n$-place decimal, 126
Number line, 103
Numbers:
  base, 4
  cardinal, 3–4
  complex, 243

Numbers (*continued*)
  composite, 84, 211–212
  concept, 3
  congruence, 220–227
  even natural, 66
  irrational, 240–243
  natural, 5
  negative, 153–160
  odd natural, 66
  ordinal, 5
  positive, 153
  prime, 83
  rational, 101–105
  real, 243
  relatively prime, 85
  whole, 6
Numerals, 7
Numerator, 104

Odd natural numbers, 66
One-to-one correspondence, 15
Opposite, 153
Ordered set, 44
Ordinal numbers, 5
Origin, 103

Parallelogram, 170
Pentagon, 170
Period:
  in positional notation, 8
  of a repeating decimal, 237
Perpendicular, 184
Place-holder, 8
Polygon, 170
Polyhedron, 176
Positional notation:
  base eight, 227–230
  base ten, 8–11
  base two, 230–234
Positive integers, 153
Prime number, 83
Product of rationals, 137–140
Product of whole number, 48–53
Proper fraction, 104

Quadrilateral, 170
Quotient:
  definition, 65
  incomplete, 71

Radius:
  of a circle, 174
  of a sphere, 178
Rational numbers, 101–162
Rational operations, 65
Rational point, 104
Ray, 103
Real numbers, 243

Reciprocal, 139
Rectangle, 170
Reflexive property:
  of congruence, 221
  of equality, 22
Regular polygon, 170
Relatively prime numbers, 85
Remainder, 71
Repeating decimal, 237
Right circular cylinder, 179
Roman numerals, 8

Separatrix, 125
Sets:
  complement of, 14–15
  disjoint, 13
  empty, 13
  equivalent, 15
  intersection of, 12
  ordered, 44
  union of, 12
Sideline of a table, 25
Sieve of Eratosthenes, 84
Sphere, 178
Square, 170
Subset, 14
Subtraction:
  borrowing, 41
  definition, 37
  techniques, 40–43, 154–157
Subtrahend, 37
Sum, 21
  of geometric series, 236
Surface area:
  rectangular box, 177
  right circular cylinder, 179
  sphere, 178

Symmetric property:
  of congruence, 221
  of equality, 22

Terminating decimals, 126
Tests for divisibility, 65–68, 203–207
Time, 182
Transitive property:
  for congruence, 221
  for equality, 22
  for inequality, 44, 161
Trapezoid, 170
Trichotomy property, 43, 161
Triangle, 170
Twin primes, 208

Union of sets, 12
Unit cube, 176
Unit square, 137

Velocity, 182
Volume:
  rectangular box, 176
  right circular cylinder, 179
  sphere, 178

Weights, 181
Well-ordering property, 44
Whole numbers, 6

Zero, 5–6, 8
  division by, 66